W0230401

Advancing Global Bioethics

Series Editors
Bert Gordijn, Dublin City University, Dublin, Dublin, Ireland
Henk A.M.J. ten Have, Pittsburgh, USA

For further volumes:
http://www.springer.com/series/10420

Leonard Tumaini Chuwa

African Indigenous Ethics in Global Bioethics

Interpreting Ubuntu

Volume 1

 Springer

Leonard Tumaini Chuwa
St. Vincent's Healthcare
Duquesne University
Pittsburgh
Pennsylvania
USA

ISSN 2212-652X   ISSN 2212-6538 (electronic)
ISBN 978-94-017-8624-9   ISBN 978-94-017-8625-6 (eBook)
DOI 10.1007/978-94-017-8625-6
Springer NewYork Heidelberg Dordrecht London

Library of Congress Control Number: 2014932537

To the heroes of Ubuntu who worked so hard to formalize Ubuntu:
Mwalimu Julius Kambarage Nyerere,
Léopold Sédar Senghor,
Kwame Nkrumah,
Nelson Mandela,
Desmond Tutu.

Preface

Ubuntu is a worldview and a way of life shared by most Africans south of Sahara. Basically Ubuntu underlines the often unrecognized role of relatedness and dependence of human individuality to other humans and the cosmos. The importance of relatedness to humanity is summarized by the two maxims of Ubuntu. The first is: *a human being is human because of other human beings*. The second maxim is an elaboration of the first. It goes; *a human being is human because of the otherness of other human beings*. John Mbiti combines those two maxims into, "I am because we are, and we are because I am." Ubuntu worldview can provide insights about relationships with communities and the world that contribute to the meaning of Global Bioethics. Ubuntu can be described as involving several distinct yet related components that can be explored in relation to major strands of discourse in contemporary Bioethics. The first component of Ubuntu deals with the tension between individual and universal rights. The second component of Ubuntu deals with concerns about the cosmic and global context of life. The third component of Ubuntu deals with the role of solidarity that unites individuals and communities.

Ubuntu has a lot in common with current discourse in bioethics. It can facilitate global bioethics. It can inspire the on-going dialogue about human dignity, human rights and the ethics that surround it. It can inspire and be inspired by global environmental concerns that threaten the biosphere and human life. Ubuntu can critique the formal bioethical principles of autonomy, justice, beneficence and non-maleficence. Above all, Ubuntu can create a basis for dialogue and mutually enlightening discourse between global bioethics and indigenous cultures. Such a dialogue helps make advancements in bioethics relevant to local indigenous cultures, thereby facilitating the acceptability and praxis of global bioethical principles.

December 2012   Leonard Tumaini Chuwa A.J., Ph.D.

Acknowledgement

I extend my deepest gratitude to Professor Gerald Magill for his challenge and inspiration into exploration of this wonderful and exciting venture. I have benefited most from his remarkable expertise and experience. I extend my thanks to Professor Henk ten Have. I benefited from Dr. ten Have's invaluable advice and constructive critique. I extend my thanks to Professor Aaron Leonard Mackler. Dr. Mackler's works on Jewish bioethics inspired me into exploring the philosophy of Ubuntu. Thanks to Rev. Dr. John Boettcher and Richard Stritter for their proofreading skills from which I profoundly benefited. Thanks to Fr. Edward Bryce, Duquesne University and John Carroll Jesuit University for their scholarship support. I likewise thank my religious community, the Apostles of Jesus, for allowing me to spend time in academic research. Last but by no means least, I thank my parents and siblings for the sacrifice they have always made in support of my desire and love for academics and research.

Contents

Chapter 1
Introduction: The Culture of Ubuntu

Bioethics is a relatively new formal academic discipline. An increasing sense of the significance of Global Bioethics is emerging in which indigenous cultures are recognized as making valuable contributions to the general field of bioethics. The culture of Ubuntu is a representative example of an indigenous African ethics that can contribute to an emerging understanding of Global Bioethics. Ubuntu, which has existed for centuries, is a sub-Sahara African culture that refers to respectful treatment of all people as sharing, caring, and living in harmony with all creation.

This book intends to bring Ubuntu ethics into the dialogue table with some of the major global bioethics. The book compares and contrasts Ubuntu with ethics of care, UNESCO Declaration on Bioethics and Human Rights, and Catholic social and ethical tradition. The scope of the book is basic groundbreaking initiative that seeks to validate Ubuntu as a dialogue partner, which seeks recognition and place at table of global bioethics. Ubuntu represents many indigenous people's ethics that have always either taken for granted or simply ignored by scholars of global bioethics. This book can neither be said to exhaust the riches of the philosophy of Ubuntu, nor its credibility as a global ethic. It is an eye opener that invites to research more into what the philosophy of Ubuntu entails.

Felix Munyaradzi articulates that according to Ubuntu culture, an ideal and meaningful life is a product of inner peace, which results from harmonious relationships among individuals, between individuals and society, and between people and their environment. This work discusses the three constituent components of the culture of Ubuntu: Human rights based on human dignity, Human cosmic context, and Social solidarity. In other words, Ubuntu is comprised of values that enable and maintain harmony among human beings, between people and their environment, and between people and the cosmos.[1] According to the ideal of Ubuntu ethics, pursuit of this harmony and tranquility in creation should occur at individual, societal and cosmic levels. Thus in the tradition of Ubuntu, cooperation between individuals, social cultures, and creation is of utmost importance.[2] The first section of this chapter explores emergence

---

[1] Murove (2004, p. 200).

[2] Richards (1980, pp. 76–77).

L. T. Chuwa, *African Indigenous Ethics in Global Bioethics,* Advancing Global Bioethics, DOI 10.1007/978-94-017-8625-6_1, © Springer Science+Business Media Dordrecht 2014

of global bioethics in relation to the culture of Ubuntu which has evolved over time but which has always been there. The second section explores Ubuntu philosophy in particular.

1.1   Emergence of Global Bioethics

For many centuries indigenous cultures such as Ubuntu have informally recognized the metaphysical ontological unity of reality. The fact that individuals have their inherent rights has always been based on the premise that individual persons are held together by the planet earth and her biosphere. Hence nature has had a special place in personal and societal existence. It is the matrix within which existence is made possible. There has always been inter-societal and trans-societal code of ethics among indigenous peoples. The Chagga of Kilimanjaro in the modern Tanzania, for example, have rules and guidelines, which guide societies and individuals in their relationships with people of other societies. Some of this code of ethics is found in their sayings, proverbs and songs. One of their maxims is, don't fight a stranger unless you have to, that is, in self-defense. Another saying translates into, this world belonged to many ancestors before us, it will belong to many generations after us.

This whole statement is contained in the saying by the Chagga of Uru which goes, *Oruka lu n'maseiyano*. The moral of the statement is to treasure the planet earth by being good stewards for it because as we enjoy it now, it has been enjoyed by many others who kept it well for us; we have a moral obligation to keep it well for future generations. This indigenous code of ethics, although shared by all Sub-Saharan peoples was never formalized since it was regarded as conventional wisdom, which was orally communicated through generations for the good of the entire human race and the planet. The following part of this chapter will explore the formal birth of global bioethics, which is actually an unconscious recognition of the indigenous perspective, which had been there for many centuries.

1.1.1   Inevitable Birth of Global Bioethics

The notion of bioethics has existed for as long as the art of medicine has existed. Formal bioethics can be traced at least as far back as at the times of Hippocrates. However this kind of bioethics was mostly medical ethics. Its focus was strictly between physicians and their patients. Among indigenous peoples all over the world, bioethics has always been part and parcel of their daily lives. The philosophy of Ubuntu in this case represents existence of informal but credible bioethics among indigenous peoples.

1.1.1.1   Limited Scope of Medical Ethics and the Increasing Need for Global Bioethics

Even though informal bioethics can be said to be as old as humanity itself, formal bioethics has been associated with medical practice and the art of healing. In other words there has been formal medical ethics prior to formal bioethics in the sense that medical ethics is a branch of bioethics. Bioethics as an academic discourse is relatively new. Arguably formal medical ethics was in existence by the time of Hippocrates between the fourth and fifth centuries before Christ. Formal bioethics, therefore, evolved from medical ethics, although informal bioethics is as old as human genre itself.[3] The birthday of formal bioethics is as late as in 1970 when the term "bioethics" was officially introduced into scholarly literature.

As an academic discourse, bioethics is, by nature, interdisciplinary and oriented into the future of humanity. Bioethics "has to emphasize that human beings are part of nature."[4] Formal bioethics was born out of necessity. Citing and supporting Potter's view ten Have argues that medical bioethics was: Too narrow to address what are, in his view, the basic and urgent ethical problems of humankind that are threatening the human survival … a new science of survival is necessary… what we currently have is medical bioethics. It needs to be combined with ecological bioethics. Both approaches in bioethics should be merged in a new synthetic approach called *global* bioethics.[5]

Strictly speaking, bioethics is universal and all-encompassing in its approach. Ten Have distinguishes "international" and "planetary" or "global" bioethics. While international bioethics concerns bioethical issues and situations that "transcend national boundaries," global bioethics "is not merely a matter of crossing borders," as "it concerns the planet as a whole."[6] Ubuntu bioethics is more universal than it is territorial, even though each peoples have always had a code of their particular bioethics. The entire universe belongs to the realm of "other" without which the "self" can never be. Care for the "other" is an ethical imperative for each human person.

1.1.1.2   Political Bases for the Genesis of Global Bioethics

There are many contributing factors that either consciously or unconsciously but necessarily contributed to the emergence of global bioethics. Socio-economic factors like slave trade, colonization, and imperialism have aroused a consciousness that necessitated emergence of global bioethics. Slavery and colonization which changed its form into imperialism could not go on for too long without provoking a reaction. It was like a pendulum that swung so much to one side necessitating swinging to the opposite direction.

---

[3] Ten Have (2013, p. 601).

[4] Ten Have (2013, pp. 603–604).

[5] Ten Have (2013, pp. 603–604).

[6] Ten Have (2013, p. 604).

World War I, for example, resulted from countries in Europe making multiple defense agreements and alliances out of suspicion and fear. The objective of such alliances was preparedness should one of such countries be attacked. Russia allied with Serbia, Germany with Austria-Hungary, France with Russia, Britain with France and Belgium, Japan with Britain. When Austria-Hungary declared war on Serbia, suddenly Russia got into the war to defend Serbia, Germany declared war on Russia, France got into the war against Germany and Austria Hungary. When Germany attacked France, Britain was involved to defend France. Being Britain's ally, Japan entered the war. Eventually Italy and the United States entered the war to defend their allies. World War I was already on as the major world nations were all involved.[7] Since Germany was the initiator of the First World War and it lost "in 1919, Lloyd George of England, Orlando of Italy, Clemenceau of France and Woodrow Wilson from the US met to discuss how Germany was to be made to pay for the damage World War I had caused." In itself, this was the initial cause of World War II.[8]

Woodrow Wilson wanted a treaty based on his 14-point plan which he believed would bring peace to Europe. Georges Clemenceau wanted revenge. He wanted to be sure that Germany could never start another war again. Lloyd George personally agreed with Wilson but knew that the British public agreed with Clemenceau. He tried to find a compromise between Wilson and Clemenceau.[9]

Since Germany was not happy with the harshness of terms of the Versailles treaty, especially because it could not pay the money due to her devastated economy, they voted to power Adolf Hitler. Hitler allied with Mussolini of Italy in the Rome-Berlin Axis Pact 1936. Hitler allied also with Japan in the Anti Comintern Pact. When Hitler ordered German troops to attack the Rhineland, the Second World War was on its way regardless of unpreparedness of the nations that had been devastated by the First World War. This situation, though unconsciously, was calling for a reaction that would prevent world wars and unhealthy alliances.[10]

Since the world was trying to find proactive way to avoid another war, the League of Nations was formed. However, the league had major flows which were seeds of its own destruction. It excluded some countries such as Germany and Russia. Germany was left out as a punishment for starting the First World War while Russia was left out of fear of communism. United Stated had had a change of government before the treaty was signed and the new Republican government opted out of the league. The league did not have power to implement its own resolutions such as stopping countries from trading with aggressive countries. The league did not have an army. Member countries were reluctant to supply soldiers mainly out of fear

---

[7] Kelley (2013).

[8] "World War Two—Causes." http://www.historyonthenet.com/WW2/causes.htm (accessed September 25, 2013).

[9] "World War Two—Causes." http://www.historyonthenet.com/WW2/causes.htm (accessed September 25, 2013).

[10] World War Two—Causes." http://www.historyonthenet.com/WW2/causes.htm (accessed September 27, 2013).

of provoking the aggressive countries that had been excluded from it. World War II Two resulted from, among other reasons, the failure of the League of Nations.[11]

The Second World War necessitated the foundation of the United Nations. Representatives of 50 countries met in San Francisco California, at the United Nations Conference on International Organization to draw up the United Nations Charter. Major players of the charter included China, The Soviet Union, The United Kingdom, and the United States. The Chatter was signed on June 26, 1945. Since Poland was not in attendance it signed the charter later and became the 51st of the original members. "The United Nations officially came into existence on October 24, 1945, when the Charter had been ratified by China, France, the Soviet Union, the United Kingdom, the United States and by a majority of other signatories."[12] The birth of the United Nations was an important step into the recognition of global bioethics.

The milestone contributor into the birth of global bioethics is the UNESCO (United Nations Educational Scientific and Cultural Organization). UNESCO was conceived in 1942 during the Second World War at the Conference of Allied Ministers of Education (CAME) in the United Kingdom. The participants intended to "develop ways to reconstruct education around the world once WWII was over. As a result, the proposal of CAME was established that focused on holding a future conference in London for the establishment of an education and cultural organization from November 1 to 16, 1945." The conference was held immediately after the First World War and was attended by 44 representatives from participating countries. The resolution of the participants was to create "an organization that would promote a culture of peace, establish an "intellectual and moral solidarity of mankind," and prevent another world war. When the conference ended on November 16, 1945, 37 of the participating countries founded UNESCO within the Constitution of UNESCO."

The UNESCO constitution came into effect on November 4, 1946. UNESCO has ever since expanded into a real global organization with 195 member states across the globe.[13] The birth of the UNESCO was in many ways the formal birth of global bioethics, especially because of its Declaration on Bioethics and Human Rights. The recognition of essential global unity of humankind and between human race and its environment, the planet earth, has always been a core belief of Ubuntu philosophy.

1.1.1.3 Demographical Conditions that Necessitated Emergence of Global Bioethics

The tremendous explosion in population growth after the Second World War resulted in an increase in life expectancy. This increase was unfortunately contradicted

---

[11] "World War Two—Causes. "http://www.historyonthenet.com/WW2/causes.htm (accessed September 25, 2013).

[12] "History of the United Nations." http://www.un.org/en/aboutun/history/ (accessed September 25, 2013).

[13] Briney (2011).

by a simultaneous increase in mortality of children under the age of 5 in the developing countries. This ethical irony was one of the eye openers, which necessitated global bioethical solidarity. World Health Organization (WHO) notes that majority of the mortality of the children in the Third World resulted from preventable diseases. WHO also notes that the unfortunate situation is being escalated by the ever widening gap between the rich and the poor; and between rich/Northern countries and poor/Southern countries.[14] The widening gap between rich and poor countries is the major underlying cause of neoliberalism. Benatar provides shocking data that reveals the already widened and ever-widening disparity between the rich and the poor. He states:

> In 1960 the richest 20% of the world's population was 30 times richer than the poorest 20%. In 1990 this rate had increased to 60 times. In the same year—that is in 1990—the debt of the poor and low-income countries was $US1.3 trillion, twice as high as in 1980, while in 1995 the debt had grown to $US1.9 trillion.[15]

This situation is not only demanding global ethical attention, it is destructive of the very nature of humanity. It compromises human dignity. From Ubuntu perspective it is utterly absurd and contrary to the very meaning of humanity.

1.1.2   UNESCO Declaration on Bioethics and Human Rights as Appropriate Response to the Needs of the Times

1.1.2.1   Globalization

One of the distinguishing phenomena of the nineteenth, twentieth and twenty-first centuries is globalization. Globalization includes, among other things, effective communication and transportation that bridges all parts of the world efficiently. Globalization, however, comes with its own drawbacks. Ten Have states that "The main source of bioethical problems is the process of globalization, particularly neoliberal market ideology." Globalization brings to global awareness ethical challenges facing other parts of the world. Ten Have lays down the environment that necessitate global bioethics when he states, "Faced with new challenges such as poverty, inequality, environmental degradation, hunger, pandemics, and organ trafficking, the bioethical discourse of empowering individuals is no longer sufficient." Hence universal bioethics is inevitable. There is clearly a need for a universal perspective and framework for bioethics "concerned with applying and implementing" the "Universal framework" of bioethics.[16]

Faced with the rapidly ever increasing globalization "the main challenge today is the impact of neoliberal market ideology worldwide." Ten Have proposes "a broader framework" such as the one "provided in the Universal Declaration on Bioethics

---

[14] World Health Organization (2009).

[15] Benatar (1998, pp. 295–300).

[16] Ten Have (2013, p. 600).

and Human Rights, presenting a wider range of ethical principles going beyond the individual perspective, including solidarity, care, social responsibility, and respect for human vulnerability." Ten Have's argument reveals a new phase in the evolution of bioethics. He posits:

> It can therefore be argued that bioethics has now entered a new phase, that is, global bio-ethics…In this new stage, global bioethics needs to go beyond the focus on human beings as autonomous individuals, emphasizing the interconnectedness of human beings, and the interrelations between human beings and the environment. This means building bridges between the present and the future, science and values, nature and culture, and human beings and nature, exactly as argued by Potter.[17]

Ten Have concludes his argument with a statement which replicates and proves the relevance of Ubuntu. He states, "The future of the human species can only be guaranteed if humanity itself is regarded as a collectivity or a 'global community'."[18]

1.1.2.2 Infectious Diseases

Infectious diseases and pandemics remind humanity of its common vulnerability and the consequent need for solidarity. Tuberculosis is an example of an infectious disease with a potential of becoming a pandemic in a short time, because it is airborne. Generally, Tuberculosis affects poor populations because of their lack of proper accommodation and their inevitable tendency to live in congested shared spaces. Conditions such as Acquired Immunodeficiency Syndrome—AIDS makes Tuberculosis even more dangerous as it takes advantage of the compromised immunity of AIDS sufferers.[19] Like Tuberculosis, Acquired Immunodeficiency Syndrome—AIDS affects poor populations more than it affects affluent populations. The poor take unreasonable risks such as prostitution, for the sake of survival. They spend their lives responding to crises since life to them ends up being mere struggle for survival. Thus, the vicious cycle of poverty and disease, which dehumanize the poor, is doubly unethical. Since the poor constitute the majority of the world population, their plight cannot be simply ignored.

Commercialization of medicine which started in the North/West, has lead into inhuman unethical exploitation of the poor South/East population. This situation cries for attention as it hampers human dignity and compromises the very essence of human nature and its essential rights, the greatest of which being the right to life. Profit maximization by global pharmaceuticals refuse to invest in drugs that are needed for treatment of diseases that affect poor populations because investing in such drugs is not good for profit. Consequently, diseases such as Tuberculosis, Typhoid, Cholera and AIDS don't get the attention they deserve from the pharmaceuticals. Thus the poor become increasingly ostracized and marginalized. The vicious cycle of poverty, exclusion and victimization escalates.

---

[17] Ten Have (2013, p. 608).

[18] Ten Have (2013, p. 611).

[19] World Health Organization (2010b, p. 3).

Contrary to the refusal to invest in diseases that affect populations of the Third World countries there is heavy investing in Cancer research because Cancer is more of a threat to the global rich populations than it is to poor populations. Clearly Ubuntu philosophy would not tolerate the kind of discrimination, which is evident in the practice of major global pharmaceuticals. World Health Organization WHO urges governments to protect their citizens from infectious diseases such as Tuberculosis. However, most governments of the Third World states lack the means to protect their populations from such diseases. Realizing Third World nations' handicap, WHO "put the responsibility on the shoulders of the community of nations". Unfortunately, WHO cannot coerce individual nations to fulfill this ethical obligation.[20] However, WHO's lack of coercive power does not change this ethical responsibility. It cannot be simply ignored. It remains always a challenge glaring at all able nations of the world.

1.1.2.3   International Trade

Unfair international trade is one of the major sources of ethical issues that call for the presence and attention of Global bioethics. One of the issues that need the attention of global bioethics is the organized and systematized dependency of poor nations on rich nations, which is perpetuated and escalated by unfair trade. Capitalistic desire for profit maximization needs to be regulated so that nobody or a state is used as a means to profit another person or state. A typical example of the exploitation of a nation by another is that seen between most Third World countries, which have been reduced to suppliers of raw materials and cheap labor rather than an equal partner in trade relationship. Although they produce raw materials, such countries usually fail to purchase the finished products, if they are exported back from the exploiting country. In itself, the status quo reveals unfairness.[21]

There are many situations in which industrialized countries use Third World countries unfairly. One good example is that used by Wynberg, Schroeder and Chennells. Developed pharmaceuticals benefit from underdeveloped peoples by taking their traditional herbal remedies, which have been proved successful over many years of experimentation. Such pharmaceuticals make drugs from the indigenous remedies, then monopolize the drugs while controlling their cost so that the indigenous people, who are the original innovators, are unable to purchase the drugs. This cruel, purely for 'profit state' of affairs is unethical and needs to be confronted since the indigenous peoples rightfully own the original right to the drug.[22]

---

[20] World Health Organization (2010b).

[21] Wynberg et al. (2009).

[22] Wynberg et al. (2009).

1.1.3 UNESCO Declaration on Bioethics and Human Rights as an Unconscious Recognition of Ubuntu

1.1.3.1 Humans should not be Used as Mere Means to Whatever End

One of the greatest values of Ubuntu is its holistic approach when it comes to dealing with fellow human beings. No bioethical principle can justify using of a human being as a means to an end. Since a human person is an integrated unity, bioethical principles should not be treated independently of each other. There has been an ironical misuse and exploitation of the bioethical principles to defend unethical practices. In such situations the gravity of the abuse is increased rather than lessened. Exaltation of the bioethical principle of autonomy over the rest of the principles seems to be the most exploited channel. The assumption in question is that as long as an individual provides fully informed consent to anything then it is not possible to consider him exploited. So to avoid exploiting a person all that is needed is to make sure that person is fully informed and that the person makes a free consent. As good as this *status quo* may seem, it can, and has been used to exploit and marginalize people. In the very attempt to avoid paternalism one may be paternalistic.[23] It is not enough to provide information and seek consent. Some acts are evil in themselves, some are evil because of the intention of the doer, and some are evil because the recipient of the act is a victim who has no choice but to consent.

There has been escalation of vicious cycle of poverty that gradually leads to more dehumanization of the poor through exploitation. The dehumanization leads to more marginalization and exclusion of the poor from the very community of humanity. A typical example is that which Eiddows rightly calls modern slavery.[24] Coerced by their poverty and need to survive, victims of trans-border prostitution have no choice but take the risk of allowing themselves be manipulated and used as pleasuring tools by people elsewhere who could pay someone else who would be considered the owner of the victim. The victim has no choice but to consent to being used. This kind of consent even if based on full information, is not free. In other words it is not consent at all. It is compulsion—the very opposite of consent. The act itself is not morally indifferent as it is exploitative and manipulative. It cannot be ethically justified.

Extraction, exportation and use of human eggs from the Third World countries for research purposes is equally abusive since it is exploitative and dehumanizing. The act is wrongful in itself and in its taking advantage of the vulnerability and neediness of the poor. Once again, the alleged consent in this situation is not *free* because the donor has no choice. The donor can rightly be viewed as a victim. There is need for global bioethical attention to such practices, which have recently escalated. Eiddows underlines the "need for global norms and protection." Underlying

---

[23] Widdows (2009, p. 6).

[24] Widdows (2009, pp. 10–16).

such practice is both gender and racial prejudice and inferiority.[25] Incapacitating the poor for the sake of forcing consent from them is a major bioethical concern. The practice thrives in the assumption of inequality of human beings which is ethically unjustifiable. The exploitative consent given by a victim in crisis who is struggling to survive is not only paternalistic; it contradicts the very concept of ethics and morals since consent presupposes freedom.

In sum, any action that takes advantage of, or uses an individual as a means for another person or any other end is essentially unethical. The UNESCO Declaration on Bioethics and Human Rights is in full agreement with Ubuntu in the sense that they both discourage reducing any human person into a means. Humans are ethically always ends in themselves.

1.1.3.2 Increasingly Obvious Need for International Bioethical Policymaking Board

Ubuntu has one loaded principle: Whatever one does to another, he or she does it to himself or herself. Human action defines the doer. Even though Ubuntu is not systematically organized the proto principal, "I am who I am because you are who you are", is a universal maxim that forbids malicious action to either human or non-human part of the universe. The statement means respect for individual autonomy, justice, beneficence, nonmaleficence and solidarity, all in one. Writing from the perspective of organized global bioethics, ten Have argues that transnational interaction and interconnectedness, environmental ethical concerns that affect individuals, populations and the planet, necessitate a global approach. His main premise is the fact that "Concern for individuals is not incompatible with concerns for the biosphere."[26] Consequently, human family should come together in addressing issues that affect the species or the planet earth. Ten Have notes that "another effect of globalization is the increasing need in global bioethics for international policymaking. The interconnected nature of ethical problems today requires international cooperation and regulation." International cooperation in bioethics is a requirement because, as ten Have notes, "Regulation at the level of the nation-state is no longer sufficient…Because of the need for international cooperation, many international organizations (WHO, UNESCO) are now active in the field of global bioethics."[27]

When dealing with a bioethical dilemmas such as that of increasing atmospheric pollution that is responsible for global warming, for example. Every individual, every society and every nation should be involved because adverse actions of one hurt all. There is need for an international and transnational policy making board that oversees and protects common good for all. Global bioethics that binds every person to the good of all people and the planet is inevitable. A good illustration of this fact is the use of chemical weapons by the Syrian president Bashar Assad in

---

[25] Widdows (2009, pp. 10–16).

[26] Ten Have (2013, p. 607).

[27] Ten Have (2013, p. 607).

August 2013, which hurt not only innocent citizens, including children and seniors in his country but which also caused a lot of anxiety and feelings of insecurity to majority of global population. It is possible for one person to destroy multitudes of people. Thus Ubuntu philosophy is demonstrably validated. In other words, Human actions may have both personal and societal/global consequences. There is need for stronger sense of global bioethics. There is even greater need for global board that will regulate unethical behaviors that may be harmful to individuals and societies in the world, regardless of national sovereignty.

1.1.3.3 The Increasing Need to Recognize Human Basic Equality Globally

One of the important premises of the possibility of ethics is recognition of basic human equality. All ethical principles are based on the assumption that human beings are essentially equal. Persons are equal in their dignity, which commands and validates human rights. Human basic equality cannot be debated. This whole truth is contained in the Ubuntu maxim, I am because you are; I am because we are. However, some humans' dignity has been severely compromised and so damaged that it hurts the entire species. The poor have been used as specimens for medical experimentation and scientific research. Using the poor that way implies that "life of a poor person is worth less than that of a wealthy person." According to Garafa, Solbakk, Vidal, et al., this dehumanizing and unethical tendency is a current practice both to poor individuals and poor nations.

During the last decade, poor and low-income countries have been increasingly involved in multicenter clinical trials aimed at expanding the fields of testing and market." It is estimated that in the year 2005, out of the "50,000 international clinical trials conducted globally, more than 40 % took place in poor and low-income countries."[28] It is unfortunate that the very champions of human rights in the world, the most civilized countries are actually perpetrators of this systemically unethical and exploitative behavior. "About one-third of 509 clinical trials sponsored by the US-based companies in 1995–2005 were conducted outside the USA, many in poor and low-income countries."[29] This exploitation is easily verified by the fact that "none of these trials have been directed towards diseases that preferentially affect the countries involved."[30] It is clearly easier said than done when it comes to the praxis of the basic ethical truth concerning human equality. Ubuntu has a lesson to teach the modern Westernized culture. Ubuntu confronts this culture by the biblical teaching: "Do unto your neighbor as you would have them do unto you."[31] Because it is others who help you be you. Others define you. The way you treat them explains to you how human you have grown to be. In sum

---

[28] Garrafa et al. (2010, pp. 500–501).

[29] Garrafa et al. (2010, p. 501).

[30] Garrafa et al. (2010, p. 501).

[31] Mathew 7:12; Luke 6:31; Leviticus 19:18; Leviticus 19:34; Tobit 4:15; Sirach 31:15; Talmud, Shabbat 31a.

Ubuntu posits humanity as one. You need to treasure your neighbor because you *are* your neighbor. Many African sayings reflect this deep, yet controversial, ethical realization. UNESCO Declaration on Bioethics and Human Rights validates Ubuntu. UNESCO formalizes and systematizes the principles of Ubuntu organizing them into a discourse.

1.2 Exploration of Ubuntu

The culture of Ubuntu presents a communal mindset for ethical decisions whereby individuals, community, and the world are connected together. Of course, the ancient culture of Ubuntu cannot articulate positions on contemporary technical bioethical issues such as brain death, genetic engineering, or cloning that are found in the developed world. Ubuntu can neither be classified as, nor compete with, the modern defined and specialized schools of thought in ethics such as modern consequentialism, deontology, or pragmatism. However, the culture of Ubuntu can provide insights about relationships with communities and the world that enable this indigenous African ethics to contribute to Global Bioethics.

1.2.1 Meaning of Ubuntu

Ubuntu is a Nguni word consisting of the "augment prefix *u-*, the abstract noun prefix *bu-*, and the noun stem *–ntu*, meaning person."[32] The word is found in most Bantu languages and shares the same construction, or same root, or same phonetics, or the same concept. Most Bantu ethnic groups use a phonological variant of the word but its meaning, word-view and application are universal to the indigenous people of Africa South of the Sahara. The Swahili people of East Africa, for example, use *Utu*, Kikuyu of Kenya use the word *Umundu*, Merians of Kenya use *Umuntu*, The Chagga of Tanzania use *Undu* and the Sukuma people of Tanzania use the word *Bumuntu*.[33]

This shared world-view, the culture of Ubuntu, is articulated by Broodryk. He describes Ubuntu as "a comprehensive ancient African worldview based on the core values of intense humanness, caring, sharing, respect, compassion and associated values, ensuring a happy and qualitative human community life in a spirit of family."[34] Asante, Miike and Yin describe the Ubuntu worldview as multidimensional, representing "the core values of African world views: respect for any human being, for human dignity and for human life, collective shared responsibility, obedi-

---

[32] Asante (2008, p. 114).

[33] Asante (2008, p. 114).

[34] Broodryk, Johann. 2002. *Ubuntu. Life lessons from Africa,* 26. Pretoria: Ubuntu School of Philosophy.

ence, humility, solidarity, caring, hospitality, interdependence, and communalism, to list but a few."[35] In Louw's words, Ubuntu "can be interpreted as both a factual description and a rule of conduct or social ethics. It both describes human being as 'being-with-others' and prescribes what 'being-with-others' should be all about."[36]

1.2.2 Ubuntu is Anthropocentric, Theocentric and Cosmocentric

African culture has been termed anthropocentric because it rarely addresses God directly. God is both transcendent and immanent. Although humans pray to God directly, they often go through intermediaries because they are believed to have better access to God. Such intermediaries would include ancestors and spirits.[37] Bujo explains this view when he notes, "One who pays heed to the dignity of the human person also pleases God, and the one who acts against the human person offends precisely this God."[38]

Another important clue into understanding Ubuntu culture is that "African ethics treats the dignity of the human person as including the dignity of the entire creation, so that the cosmic dimension is one of its basic components."[39] Consequently, African ethics can only be properly understood from the perspective of being anthropocentric, societal, cosmic and theocentric. Its objective is "fundamentally life itself."[40]

Each member of the community and the community as a whole "must guarantee the promotion and protection of life by specifying or ordaining ethics and morality."[41] Life is the highest principle of ethical conduct.[42] Whatever is against life is unethical; whatever favors life is ethical. Although human life is the center of all life on earth, all life is sacred since all life is considered interdependent.

1.2.2.1 Interdependence

Ubuntu observes a network of interdependence and relationships that are divinely ordained to promote, sustain and foster life. A human person can neither be defined nor survive if separated from the society and the cosmos that enables that person's existence. It is a matter of justice to care for other humans, other lives and the non-

---

[35] Asante (2008, p. 114).

[36] Louw (2007).

[37] Mbiti, John S. 1969. *African religions and philosophy,* 16, 28. 2nd ed. New Hampshire: Heinemann Educational Books Inc.

[38] Bujo (2001, p. 2). Bujo was citing Reese-Schaffer W. 1994. *Was ist Kommunitarianismus?* Frankfurt a. M./New York.

[39] Bujo (2001, p. 2).

[40] Bujo (2001, p. 2).

[41] Bujo (2001, p. 2).

[42] Bujo (2001, p. 3).

living part of the cosmos. Len Holdstock describes this African perspective of reality as holistic. He notes that for an African, everything belongs together; humans
and the world around them belong together. Causing harm to the environment is
hurting oneself. Humans are perceived as a vital force which is interrelated with,
and contingent upon other vital forces around them. Existence is inconceivable independent of the beings' interdependence and interrelationships.[43] To explain this
mindset among the peoples of sub-Sahara Africa, Donna Richards writes that "The
traditional African view of the universe is as a spiritual whole in which all beings
are organically interrelated and interdependent...the cosmos is sacred and cannot
be objectified.

Nature is spirit, not to be exploited...All beings exist in reciprocal relationship to
one another."[44] However, Richards notes that there is tension in reality which under-
lines individual self-determination without negating the ideal of harmony in reality.
She also notes the same interdependence between spirit and matter. She states that
"The mode of harmony which prevails does not preclude the ability to struggle.
Spirit is primary, yet manifested in material being."[45]

According to Ubuntu, reality is unity in which God is both imminent and transcendent. Ubuntu may be understood by considering the philosophy of Pierre Teilhard de Chardin. De Chardin transcends the traditional dualistic and dichotomous
worldview in his powerful perception of reality as essentially one. He construes
reality neither as a division of matter and energy, nor spirit and matter but as a
single reality in the shape of two phenomena. Matter and energy are two aspects or
phenomena of a single being; there is inseparable interconnectedness and interdependence. Matter can be reduced to energy and energy to matter.[46]

Such worldview, which is based on scientific discoveries, best describes the perspective of Ubuntu. In Ubuntu the physical and the spiritual, the living and the
non-living, the human and the non-human are perceived as necessary in sustenance
of human life. Human life comes from, and is sustained by both organic and inorganic cosmos. For the sake of harmony, which is an ethical ideal, humans must treat
each being fairly according to its moral status and claim. Thaddeus Metz argues
that compared to holist and individualist conceptions of moral status, this African
"underexplored modal-relational perspective does a better job of accounting for
degrees of moral status."[47] Interdependence and relationality is the kernel of the
argument for the moral status of non-human beings.

In sum, the human person is an organism within a bigger organism, the society.
Human society is a part of the biosphere and the cosmos. God is both transcendent
and imminent in the sense that he pervades reality while at the same time remains
separate from it. Somé observes, "The close relationship between people and place

---

[43] Holdstock (2000, pp. 162–181).

[44] Richards (1980, pp. 76–77).

[45] Richards (1980, pp. 76–77).

[46] Teilhard de Chardin, Pierre. 1969. *Human energy* (Trans. J. M. Cohen), 113–162. New York:
Harcourt Brace Jovanovich.

[47] Metz (2011).

is symbolized by the bond that indigenous people recognize between a person and his or her place of birth, and also in the fact that any ritual that is performed is viewed as being tied to the geography where it takes place."[48] Being a comprehensive world view, therefore, Ubuntu is a loaded term which is defined in a variety of ways. Whichever way Ubuntu is defined, it "reveals African culture and tradition, beliefs and customs and value systems."[49] Ubuntu is the bond that underlies the cultural diversity and various value systems of most African ethnic groups south of Sahara.

1.2.2.2 Need for Otherness

In Ubuntu ethics, the importance of other persons for any human being cannot be overstated. A person is both ontologically and socially a product of other persons. Ubuntu is based on two maxims. The first is "a person is a person through other persons."[50] The second maxim is a translation of the same statement that underlines the need for diversity and plurality. It states, "A human being is a human being through the otherness of other human beings."[51] Otherness includes human diversity of languages, histories, values and customs, all of which constitute human society.[52]

Ubuntu recognizes the fact that an individual can only become conscious of his/her existence along with its rights as well as obligations towards the self, other persons and the universe by the medium of the presence of others. In other words, cut off from all others, no individual personal life is possible, let alone personal consciousness.[53] Such personal consciousness is based, not only on the living members of the society; it is based on all those who have died, from whom the present members descended. The culture of Ubuntu recognizes that present generation is a product of past generations. Many past generations have paved the way and made it possible for any current generation to be what it is now. Current generations, so to say, stand on the shoulders of past generations.[54]

Because of its 'other-oriented' worldview Ubuntu is communitarian. Mbiti express this interconnection between individuals in praxis when he states that in the Ubuntu culture whatever happens to the individual happens not just to that individual but to the community in which the individual is a member. Likewise, whatever happens to the community impacts each member of the community. When an individual rejoices "he rejoices not alone but with his kinsmen, his neighbors and his relatives whether dead or living... The individual can only say, 'I am, because

---

[48] Some (1998, p. 38).

[49] Broodryk (1997, p. 26).

[50] Shutte (1993 p. 46).

[51] Willie and Marwe (1996, pp. 1–3).

[52] Willie and Marwe (1996, pp. 2–3).

[53] Mbiti, John. 1970. *African religions and philosophies,* 141. New York: Anchor Books.

[54] Bénézet (2001, p. 27).

we are; and since we are, therefore, I am.'"[55] Gyekye notes that most Bantu languages, acknowledge that a person is, "inherently a communal being, embedded in a context of social relationships and interdependence, and never as an isolated, atomic individual."[56] No individual survival or realization is possible independent of the community.

Since the community enables individuation and its basic rights, duties and obligations, the individual owes the community—just as the community owes the individual. Neither of the two survives without the other. The community is a product of its many individuals, just as the individual is a product of many members of the community. The interdependent mutuality between the community and its members can neither be denied nor overstated. In his *African Traditions and Religions* basing his argument on a research that he made in sub-Sahara Africa, Mbiti explores the symbiotic relationship between sub-Sahara Africans and their respective ethnic communities. He notes that individual existence is only possible within corporate existence. Consequently any particular individual is simply "part of the whole." Separation from the community is not only impossible, it is inconceivable.

It is an essential duty of the community, therefore, to "make, create, or produce the individual; for the individual depends on the corporate group. Physical birth is not enough: the child must go through a rite of incorporation so that it becomes fully integrated into the entire society."[57] The role of the corporate community in the constant on-going creation of the individual commands reciprocity in form of individual cooperation in the life of the community. Among the Bantu peoples of Africa, life is characterized by constant initiation and incorporation which is expressed in rich symbolism of new birth. Such initiations are stages through which the community creates the individual, enabling him to self-realize. They are forms of incorporation into the wider world. According to Mbiti, the initiation goes on even after death. The final stage of initiation for the person "is reached when he dies and even then he is ritually incorporated into the wider family of both the dead and the living."[58]

The phrase "being with others" in Ubuntu is of central importance. It is not limited to human beings. It includes the biosphere and the cosmos, since human action affects both humans and non-human universe. Human beings are not only dependent on one another; they are dependent on the biosphere and the cosmos. Human existence is rooted in, facilitated by, and constantly related to the biosphere and the cosmos. Just as it is impossible to envision human personhood independent of the society, so it is impossible to envision human society independent of its context, the cosmos. In the words of Somé, "The natural world is an integral part of an indigenous community. Village people envision the community as including

[55] Mbiti (1970, p. 141).

[56] Gyekye, Kwame. 2002. Person and Community in African Thought. In *Philosophy from Africa. A text with readings*, ed. P. H. Coetzee and A. P. J. Roux, 319. Cape Town: Oxford University Press Southern Africa.

[57] Mbiti, John. 1969. *African traditional religions and philosophy*, 106.

[58] Mbiti, John. 1969. *African traditional religions and philosophy*, 106.

the geography and the natural world that surrounds and contains the people."[59] The peoples of sub-Sahara Africa perceive society, the biosphere and the cosmos as an extension of the self. Consequently "being with others" is necessary for personal happiness, peace, integrity and self-realization. Thaddeus Metz posits that relationality is "at the core of morality."[60]

The culture of Ubuntu is both theocentric and anthropocentric. Human beings are the center of created reality. However, humans' centrality does not preclude God's supreme sovereignty. Ubuntu holds that all reality is under God's control. Kasanane notes that traditional Africans believe that God controls the universe, human interactions and relationships among themselves, as well as with the cosmos by means of a number of subsidiaries. The "Supreme Being is interested in the way people relate to one another. A number of taboos, regulations and prohibitions exist in every society to ensure mutual coexistence."[61]

1.2.2.3 Ubuntu and Unity

Due to its steadfast belief in the importance of unity as a fundamental value, Ubuntu culture is not interested in separating, defining, and distinguishing. Kasanane expresses this worldview in his work, *Ethics in African Theology* when he says, "in African religions there is no separation between religion and ethics, between one's beliefs and one's actions towards others. Ethics is an integral part of religion."[62] Mbiti notes that religion is part and parcel of the life-style and all activities of traditional Africans. In Mbiti's words, "Because traditional religions permeate all the departments of life, there is no formal distinction between the sacred and the secular, between the religious and non-religious, between the spiritual and the material areas of life. Wherever the African is, there is his religion."[63] Whether a traditional African is sowing, harvesting in a party, in a funeral ceremony or at war, he/she is religious. Religion cannot be separated from the believer. Thus, the daily normal activities of people are at the same time acts of worship.[64] Bujo summarizes sub-Saharan holism when he writes; "no dichotomy exists in Black Africa between body and soul, or between theory and praxis—or in the present instance between the body and knowledge."[65] Reality or existence is a function of unity.

Human unity is crucial in the comprehension of existence itself. Unity is of ontological, societal, ethical and religious importance. Exploring African culture as compared to western culture, Steve Biko States: "We regard our living together not as an unfortunate mishap warranting endless competition among us but as a deliber-

---

[59] Some (1998, p. 38).

[60] Metz (2011).

[61] Kasanene (1994, p. 140).

[62] Kasanene (1994, p. 140).

[63] Mbiti (1970, p. 1).

[64] Mbiti (1970, p. 1).

[65] Bujo (2001, p. 26).

ate act of God to make us a community of brothers and sisters jointly involved in the quest for a composite answer to the varied problems of life." Biko then evaluates an African action as communal as distinguished from individualistic approach to human action. He states, "our action is usually joint community oriented action rather than the individualism which is the hallmark of the capitalist approach."[66]

The corporate view of a human person among sub-Sahara Africans is so deep that, as Menkiti relates, a person cannot be defined, as is the case in the western world, according to his "physical or psychological characteristic of a lone individual." A person transcends such criteria. A definition of a person which excludes the community falls short of necessary components that define personhood, the most important of which concern his relationality, need for community or society, and unity.[67]

Personal and societal need for the biosphere and the cosmos can hardly be overstated in the Ubuntu perspective. Although no animal has equal moral status to a human person, nevertheless, to the degree that humans relate with a particular member of the biosphere, such being has a degree of moral status that ought to be recognized and respected, especially in relation to human dignity and the need of humans for other beings.[68] Metz argues at length for the validity and plausibility of this African perspective as a theory of moral status. Metz contends that "in light of the African theory's ability to account for many widely shared intuitions, it warrants no less attention than individualist and holistic accounts as a promising form of monism."[69] The unity that Ubuntu advocates for is based on human dignity. This dignity demands human rights without excluding the rights of the biosphere, since human beings cannot be extricated from the biosphere. Consequently, Metz concludes that Ubuntu is a form of monism.

1.2.3   Ubuntu Ethics of Immortality

Due to its reverence for life, Ubuntu ethics' objective is not only preservation of the ontological life on earth but also its survival after physical death. According to Ubuntu, human life is so central, so dignified, unrepeatable, sacred and unique that it should survive physical death. Strictly, from Ubuntu perspective human life does not end. Thus, death is yet another stage of initiation in the human life's process of continual and immortal initiation.

1.2.3.1   Personal Immortality

Most African indigenous societies believe that personal immortality is achieved in two ways. On the level of an individual, personal immortality is "externalized in the physical continuation of the individual through procreation, so that the children

---

[66] Biko (2004, p. 46).
[67] Menkiti (1984, pp.170–175).
[68] Metz (2011).
[69] Metz (2011).

bear the traits of their parents or progenitors."[70] Immortality is crucial for life meaning both for the deceased and the survivors. Mbiti explains that "from the point of view of the survivors, personal immortality is expressed in acts like respecting the departed, giving bits of food to them, pouring out libation and carrying out instructions given by them either while they lived or when they appear."[71] This is the second way immortality is achieved; that is, as memory in the minds of the survivors. Such memory is maintained and treasured. When the memory fades with passage of time it is still considered present but compromised by our finite ability to remember.

It is important to note that from the perspective of sub-Sahara Africans, the *living-dead* are really present, although they have been initiated in a higher form of existence. They oversee behavior of the living, and can punish in case of immorality. "The acts of pouring libation (of beer, milk or water), or giving portions of food to the living-dead, are symbols of communion, fellowship and remembrance. They are the mystical ties that bind the living-dead to their surviving relatives."[72]

This mystical tie between the physically living and the "living-dead" is, on the part of the survivors, an obligation. The dead are kept in memory for as long as possible. When there is no one living who remembers them they are believed to have gone through other initiations into the world of spirits which is further removed from the world of the living but the ties always remain.

The "living-dead" keep moving into a dimension of time that Mbiti calls *zamani* (Swahili word for past). Such a dimension is, for most Africans, more ontologically real than future. According to Mbiti's observation, sub-Sahara Africans' time has two major dimensions: *zamani* and *Sasa* (Swahili word for now or present). The future is not real since it has not been realized, that is, it is not yet present. Since the dead are real, they are actually more real than the present since they have gone through more initiations into reality than the living.[73] Thus, even though it is important to survive death by procreation on an individual level and by personal memory on the level of the survivors, death does not end human life.

Ubuntu healthcare for the terminally ill and the dying is rich with meaning and symbolism. The whole community participates in this significant initiation of that member into the community of the living-dead. The community accompanies the dying, giving them, as Bujo notes, "the feeling and the awareness that they are included in the *process of personal growth* even as their physical strength declines… the sick and the dying find fresh courage and learn to face suffering and death with greater human dignity."[74]

This positive perspective on death and the participation of the community in the process is a great help not only to the dying but also to the living. It is both ascertaining death with dignity and a healthcare lesson for the community. The living

---

[70] Mbiti, John. 1970. *African religions and philosophies*, 1.

[71] Mbiti, John. 1970. *African religions and philosophies*, 25.

[72] Mbiti, John. 1970. *African religions and philosophies*, 25.

[73] Mbiti, John. 1970. *African Religions and Philosophies*, 22–28.

[74] Bujo (2001, p. 89).

members of the community learn to prepare and to go through this inevitable natural initiation with courage when their turn comes.[75]

1.2.3.2 The Importance of Marriage and Procreation

Due to the importance of immortality both for the diseased and the surviving society, marriage and procreation is of utmost importance. Life revolves around it. According to Mbiti traditional African marriage "is a complex affair with economic, social and religious aspects which often overlap so firmly that they cannot be separated from one another."[76] The importance of marriage in African traditional society is based on the fact that it is the central source of societal and personal immortality. Marriage is "the point where all the members of a given community meet: the departed, the living, and those yet to be born.

All the dimensions of time meet here and the whole drama of history is repeated, renewed and revitalized."[77] The centrality of marriage rests on the fact that it ascertains continuity of life and the community. Consequently, marriage is neither a personal decision nor a private matter. The community naturally expects everybody to marry both for personal immortality and for the sake of the community. It is a duty and an obligation. Bujo writes that even "sexuality is not a private matter. The goal of sexuality is to keep together the community entrusted to us by our ancestors and to bestow ever new life on this community."[78]

Prior to the advent of Christianity, sub-Sahara Africa had never considered celibacy as a valid option. Celibacy has always been a sign of selfishness, withdrawal from the community and its rhythm, and an offence against the natural law of generation and nurturing life. Benezet Bujo compares a celibate person with a magician in their action against life.[79] Just as a magician destroys life, a celibate person passively by his omission is against life. Prostitution as such or sex for mere pleasure was seldom heard of in traditional African society, mainly because of the understanding that sexual intercourse is meant for generation of life.[80]

Due to its deep rooted communitarian life, and its understanding of sexuality as a means to procreation, sub-Sahara Africa has had few cases of openly known homosexuality. Once again Bujo attributes this situation to African communitarianism. Bujo states that "one is a human being only in the duality of man and woman, and this bipolarity generates the triad man-woman-child, which leads to full

---

[75] Bujo (2001, p. 89).

[76] Mbiti, John. 1971. *African traditional religions and philosophy,* 13. London: Heinemann.

[77] Mbiti, John. 1971. *African traditional religions and philosophy,* 13.

[78] Bujo (2001, p. 37).

[79] The word "magician" in this work means a person who uses evil forces to hurt or destroy life. This use is different from the western understanding of magic. Anything that disrupts life, unity or harmony is considered evil. Magic in this context is always evil as it is intentional causation of evil. Witch doctors and medicine men/women work against magicians and magic.

[80] Bujo (2001, p. 7).

community."[81] Community being central in Ubuntu ethics because of its necessity in support of all human life, marriage between man and woman is the very first stage in the creation of larger human community and in generation of life.

Man to woman marriage is essential for survival of human community and for the sacred duty to generate and maximize life. Bujo observes that "a man-man or woman-woman relationship would not only be looked on as an egotistic isolationism which dares not take the step to full human existence; [since existence cannot be separated from the duty to generate, protect and maximize life] it also leads to a sexist discrimination against part of the human race and shows an unwillingness to accept the enrichment that comes from heterogeneity."[82]

In case a couple could not procreate, the community improvise a way of helping childless couple participate in the life of the community by adoption of children of relatives, generating children for an infertile husband via his siblings and polygamy. Homosexuality is against real community, thus against life and human race. It is always considered evil and of great immorality.

In most African traditional societies, failure to give birth to a child is equivalent to death. Although the society has had a myriad of remedies for a childless couple, even if extended family is such that siblings' children are one's children, having one's own child is the norm and ideal for which all human beings should strive. Mbiti states this fact more explicitly when he writes, "Unhappy is the woman who fails to get children, for whatever other qualities she might possess, her failure to bear children is worse than committing genocide: she becomes the dead end of human life, not only for the genealogical line but also for herself"[83], since to be truly alive is to be a link in the chain that receives life and passes it on via procreation.

Being part of the chain of life earns an individual personal immortality. Mbiti writes that the greatest plight of a childless wife is the fact that "when she dies, there will be nobody of her own immediate blood to remember her, to keep her in the state of personal immortality: she will simply be 'forgotten...her husband may remedy the situation by raising children with another wife; but the childless wife bears a scar which nothing can erase."[84] As much as Ubuntu protects every individual, assuring social and psychological security to all society members, a childless wife cannot be protected. Because of her fate, her family and the society suffer along with her.[85]

In sum, marriage depicts the traditional African worldview in practically all its perspectives. African perspective on marriage is a microcosm of the communitarian understanding of the Ubuntu worldview. Such worldview is represented in the symbolism, rituals, songs, proverbs, stories and poems used in the celebration of marriage.

---

[81] Bujo (2001, p. 6).

[82] Bujo (2001, p. 6). The words in the brackets are mine.

[83] Mbiti, John. 1969. *African traditional religions and philosophy,* 107.

[84] Mbiti, John. 1969. *African traditional religions and philosophy,* 107.

[85] Mbiti, John. 1969. African traditional religions and philosophy, 107–108.

1.2.3.3 Ubuntu Theory of Moral Development

Due to its respect for life, the culture of Ubuntu reverences the process of birth. Most members of the society are involved in the process. However, the process does not end with physical child birth. A pregnant woman's labor pain is gradually taken over by the immediate family, the extended family and eventually the community. From the moment of conception a child starts growing more into becoming a child of the society rather than of its immediate parents. The life-long process of initiation that Mbiti refers to and its many rites and rituals are geared towards such incorporation.

Thus the community gradually takes over the process of helping the child realize itself. The physical placenta and umbilical cord represent "separation of the child from the mother, but this separation is not final since the two are still near each other. But the child begins to belong to the wider circle of society."[86] The community has to help the child become a human being, a person. In Mbiti's words "nature brings the child into the world, but society creates the child into a social being, a corporate person."[87]

The community has in place a continual process of formation of a child into a mature person in the community. Such process is usually in the form of initiation and societal incorporation. The child is helped into self-realization by the society. Self-realization, however, is always accomplished through the medium of the community. Shutte makes it clear that, "Our deepest moral obligation is to become fully human. And this means entering more and more deeply into community with others. So although the goal is personal fulfillment, selfishness is excluded."[88] This growth is essential not only because it is the path to acquisition of membership in the society but also because it is constitutive of the essence of humanness. A person who fails to grow into relating with other persons in an acceptable way is regarded as inhuman.

Most Bantu languages use phrases like "he is not a person" or "he is an animal." Swahili, which is used as a national language in Tanzania and which is spoken in some East and Central African countries, uses the phrase *hana utu* (s/he lacks personhood).[89] Thus, it is imperative that the society recognizes one as a person, therefore a member. Lack of personhood means, at the same time, lack of the essential qualifications to belong to the society, thus incapable of membership.

Every person is helped by the community and should cooperate in his own process of moral development. Essential in the process of moral development is becoming a part of the community as a whole. Gbadegesin refers to this objective and expectation when he writes "every member is expected to consider him/herself

[86] Mbiti, John. 1969. *African traditional religions and philosophy,* 110.

[87] Mbiti, John. 1969. African traditional religions and philosophy, 107.

[88] Shutte, Augustine. 2001. *Ubuntu: An ethic for the new South Africa,* 30. Cape Town: Cluster Publications.

[89] Bhengu (1996, p. 27).

an integral part of the whole and to play an appropriate role towards achieving the good of all."[90] Doing that, however involves reciprocity, since a person can only self-realize through other persons. This observation led Metz and Gaie to conclude that "sub-Saharan morality is essentially relational in a way that other Western approaches usually are not."[91]

Metz and Gaie compare the sub-Saharan sense of justice, impartiality, and understanding of human rights with the modern Western perspective. The basic substance of the human rights argument and theory is contained in the sub-Saharan understanding of justice. Metz and Gaie observe that Ubuntu includes "an impartial element, part of which is a matter of individual rights. Traditional African societies have often thought of human life as having a dignity that implies recognition of certain universal human rights."[92] Human rights are not negotiable in sub-Saharan Africa. They are a given and are almost identical to the modern western conception of human rights. Metz and Gaie argue for this conclusion when they state that, "Despite the moral prominence given to their own community, indigenous sub-Saharan societies are well-known for having welcomed a stranger to their villages, giving him food and shelter for at least a period of time."[93] This practice is not charity. It is based on the understanding of human dignity. This dignity is shared by all who share human life. Metz and Gaie observe that Africans "hardly considered a foreigner outside the bounds of moral consideration and, instead, tended to view all humans as potential parts of an ideal family."[94]

There is no doubt whatsoever that the sub-Saharan indigenous communities recognized human dignity and the basic rights (human rights) due to such dignity. Such deep-rooted recognition of basic human rights is often referred to by modern sub-Saharan judicial systems. Metz cites some remarks by the South African constitutional court which occasionally appeals to Ubuntu and its understanding of basic human rights when making legal deliberations. A concrete example for this argument is found in the work of Justice Yvonne Mokgoro where he remarked "Human rights derive from the inherent dignity of the human person. This, in my view, is not different from what the Spirit of Ubuntu embraces."[95]

Actually Metz and Gaie contend that sub-Saharan indigenous Africans' concept of justice which is represented in Ubuntu can be reduced to, and transcends both Kohlberg's theory of justice (respect for equal rights of persons model) and ethics of care's perspective (relationality and reciprocity of care model).[96] The kernel of sub-Saharan morality, which is represented by Ubuntu, is human life. The principles

---

[90] Gbadegesin (1991, p. 65).

[91] Metz and Gaie (2010, p. 275).

[92] Metz and Gaie (2010, p. 275).

[93] Metz and Gaie (2010, p. 283).

[94] Metz and Gaie (2010, p. 283).

[95] Justice Yvonne Mokgoro of the Constitutional Court of South Africa, *The State versus T. Makwanyane and m Mchumu*, para. 309. Cited by Metz (2007, p. 329).

[96] Metz and Gaie (2010, p. 283).

of justice and care, even Ubuntu relationality are a means to an end, which is, maximization of quantity and quality of human life.

Onah states this objective of African morality bluntly when he writes: "at the center of traditional African morality is human life. Africans have a sacred reverence for life…to protect and nurture their lives, all human beings are inserted within a given community." Community therefore is a means to an end: human life. Onah notes that "the promotion of life is therefore the determinant principle of African traditional morality and its promotion is guaranteed only in the community. Living harmoniously within a community is therefore a moral obligation ordained by God for the promotion of life."[97]

The community and each of its members have to participate in the duty of promoting and preserving life since, human life, in itself, commands human rights due to its inherent dignity. Thaddeus Metz coins a moral principle from this unanimously accepted sub-Saharan African stance towards life. "An action is right just insofar as it promotes the well-being of others without violating their rights; an action is wrong to the extent that it either violates rights or fails to enhance the welfare of one's fellows without violating rights."[98]

Consequently, even if it is not always explicitly stated, it can safely be concluded that the ultimate objective of Ubuntu is protection of basic human rights. Desmond Tutu mentions values which ascertain protection of life when he writes, "Harmony, friendliness, community are great goods. Social harmony is for us the *summumbonum*—the greatest good."[99] Although Tutu does not mention that social harmony is actually a means to an end, which is human life, he mentions the vices that should be avoided because they either threaten or undermine human life. "Anything that subverts or undermines this sought-after good is to be avoided like the plague. Anger, resentment, lust for revenge, even success through aggressive competitiveness, are corrosive of this good."[100] Social harmony is the rightful context for flourishing of human life; that is why even aggressive competitiveness is perceived as a vice, rather than a virtue.

Ubuntu model of moral development includes: respect for dignity of other human beings, recognition of their personhood, establishment of human relationship with others while, at the same time, safeguarding essential respect and praxis of human rights and execution of justice. Its major objective is provision of the optimum context and environment for maximization of quantity and quality of human life. With regard to its objective—human life, Ubuntu model, which combines care and justice, is worth exploring further, especially because it is comprehensive in approach, comprising and encompassing the modern piece-meal approaches.

---

[97] Onah (2012).

[98] Metz (2007, p. 330).

[99] Tutu (1999, p. 35).

[100] Tutu (1999, p. 35).

1.3   Relevance of Ubuntu Worldview

Being a product of many centuries of human existence in relation to nature, the culture of Ubuntu is discovered or spontaneously observed rather than invented. Individuals find themselves already bonded with each other and with the cosmos as a matter of necessity. The rationale for this bonding together is a product of many centuries of cumulative experience-based survival wisdom. Such wisdom is passed on by ancestors via elders.[101]

Human identity and personhood is impossible independent of community. Individuation occurs in a dynamic relationship with others, without which there can be no human personal existence. This important truth is best explained by Benhabib in her work, *Situating the Self: Gender, Community, and Post Modernism in Contemporary Ethics*: "Because the identity of the self is inter-subjective, the 'I' can only become an 'I' in the context of a 'we.' Individuation does not precede association; rather it is the kind of associations which we inhabit that define the kinds of individuals we become."[102] Thus Ubuntu is essentially communitarian. Individuality and personhood are facilitated by context, which comprises human society, biosphere, and the cosmos. The society precedes and defines its constituents. Human action should therefore proceed from such background and context.

1.3.1   *Ubuntu Existential-Relational Epistemology*

Charles Taylor contends that our modern sense of self is constituted by some sense of "inwardness." Taylor notes that "in our language of self-understanding, the opposition 'inside-out' plays an important role. We think of our thoughts, ideas or feelings as being 'within' us, while the objects in the world which these mental states bear on are 'without.'" Taylor then contends that this sort of "localization is not a universal one … it is a function of a historically limited mode of self-interpretation, one which has become dominant in the modern West … but which had beginning in time and space and may have an end."[103]

Taylor then laments that that we are constantly losing the obvious fact that "being a self is inseparable from existing in a space of moral issues, to do with identity and how one ought to be."[104] Then Taylor argues both for neutrality in the world of morals without denying one the right to hold a position in it. He argues that being a self in the world of morals is "being able to find one's standpoint in this space, being able to occupy, to *be* a perspective in it."[105] Being a perspective in the space of moral issues does not and should not change the real human moral universals.

---

[101] Bujo (1992, pp. 21–26).

[102] Benhabib (1997, p. 73).

[103] Taylor (1989, p. 111).

[104] Taylor (1989, p. 112).

[105] Taylor (1989, p. 112).

Taylor points out how hard it is to extract the real universals from different perspectives. He states that "The really difficult thing is distinguishing the human universals from the historical constellations and not eliding the second into the first so that our particular way seems somehow inescapable for humans as such, as we are always tempted to do."[106] Following Taylor's observation and insight, there is need also for ethicists to be ethical in establishing what is considered universal for all humans, especially with regards to morals. In other words, it is possible to be unethical in the very act of doing what we consider authentic ethics. There is inescapable need to respect and pay serious attention to other ethics, or ways of doing ethics.

Ubuntu world-view affirms, challenges and inverts the platonic world of ideas, Descartes' disengaged self in *cogito ergo sum*, Augustine's doctrine of finding God from within oneself and in the order of creation and Locke's punctual self. Ubuntu provides a contrast that is necessary for a fair perspective on reality; that is, the role of otherness in ontology, theology and epistemology. Ubuntu states that the inner-self or selfhood can only be accessed via objectification; objectification, however, demands otherness. Any relationship with God or the self implies accepting another—even if by imagination. Imagination and ideas, however, result from reality of existence of other beings.[107]

Ubuntu presents a sharp contrast to the Cartesian individualist proof of existence *cogito ergo sum* (I think, therefore I am) by its communitarian existential and relational *cognatus sum, ergo sumus* (I am known, therefore we are).[108] This epistemology is relational. It means that the act of conceptualization is at its core relational, as it must involve two beings to be practicable. Moreover, present human individual action is a product of many centuries of experience, evolution and its resultant cumulative wisdom. Bujo notes that Joseph Ratzinger (Pope Benedict XVI) interestingly arrives to the same conclusion via his Christology. Ratzinger writes: "Christian faith does not find its starting-point in the atomized individual, but comes from the knowledge that the merely individual person does not exist." Ratzinger then affirms the Ubuntu understanding of a human being as a link in a long chain of evolving human history within the cosmos as its necessary context. Ratzinger states, "The human person is himself only in an orientation to the totality of humanity, of history, and of the cosmos. This is an appropriate and essential dimension of the human person as 'spirit in a body.'"[109]

Essentially, Ratzinger argues that a human person cannot exist as a monad. Solitary human existence is self-contradictory. To be human means to simultaneously be a member of human community, to actively participate as member of the present human community, and to take one's place in the on-going chain of human history which must be passed on to future generations.[110] This approach to meaning

---

[106] Taylor (1989, p. 112).

[107] Taylor (1989, pp. 111–176).

[108] See Reese-Schaffer, *Was ist Kommunitarianismus?* In Bujo (2001, p. 4).

[109] Ratzinger, Joseph. 1968. Einfuhrung in das Christentum: Vorlesungen uber das Apostolische Glaubensbekenntnis (Munich),176. See Bujo (2001, p. 4).

[110] Bujo (2001, p. 4). In this passage Bujo dwells on Franz von Baader's criticism of Descartes. Joseph Ratzinger agrees with Baader's line of argument and validates *cognatus sum, ergo sumus*

and significance of human life is shared both by the Christian social teaching and Ubuntu. Bujo contends that the principle *cognatus sum ergo sumus* (I am known, therefore we are) is superior to the Cartesian *cogito ergo sum* (I think, therefore I am) because it transcends the metaphor of knowledge. *Cognatus sum ergo sumus* "is not only a given; it is existential to such a degree that refusal to accept it must lead to the death, not only of the individual but even of the community itself."[111]

Unlike modern trends in ontology and epistemology Ubuntu recognizes the significance of healthy relationality between humans and the cosmos; between the living and the dead; and between material world and spiritual world. Bujo regards relatedness as "the decisive issue…It signifies merely an openness that goes beyond what is present and visible in a given situation."[112] Relatedness does not only facilitate individuation and intelligibility of reality, it is the kernel that keeps unity of reality as its very existence. As for human relationality, Bujo summarizes the importance of relationality between individuals when he writes "individuals live only thanks to the community"[113] which provides for the possibility of establishing relationships. By means of analogy Ruch states that an African "does not feel himself like a swimmer in a hostile and foreign sea: he is part of this sea, he participates in it as it participates in him."[114] This analogy describes both Ubuntu ontology and epistemology in a very profound way. It is a summary of the Ubuntu worldview.

Human need for community as the kernel of Ubuntu ethics has been expressed by a number of Ubuntu scholars. John Mbiti, for example, states that Ubuntu ethics is based on the premise that an individual becomes conscious of his own existence, rights, duties and obligations through other individuals, society and the environment. There is no real personal existence independent of the society and environment. Whatever affects one individual affects the entire society and its environment. Likewise, whatever affects the society affects each individual in it and that person's environment. Thus "an individual can only say, 'I am, because we are; and since we are, therefore, I am.'"[115]

1.3.2 Ubuntu Relational and Holistic Perspective on Human Disease

Bujo observed that in African traditional society, disease and illness that befell an individual always indicated that something is wrong in human relationships. Consequently, in diagnosing an individual's illness "the patient's family relationships

---

(I am thought, therefore we are) against Descartes *Cogito ergo sum* (I think, therefore I am). Ratzinger argues that "it is only on the basis of his being known that the knowledge of the human person and his person himself can be understood," In *Einfuhrung in das Christentum*, 177.

[111] Bujo (2001, p. 5).

[112] Bujo (2001, p. 3).

[113] Bujo (2001, p. 3).

[114] Ruch (1975, p. 2).

[115] Mbiti (1970, p. 141).

are studied and past conflicts interpreted anew. ... The healing rituals and witch executions in their different ways restore, or attempt to restore, harmonious social life."[116] Any attempt to effect healing cannot ignore human and environmental relationships. Diagnosis of the ailment and its causality cannot be done independent of patient's relationships with human and non-human environment. One's ailment may as well be a consequence of disturbed or non-harmonious relationships with the world of the dead."[117] Thus, disease does not always result from physiogenic causes. Many diseases result from psychogenic, spiritual and sociological causes. It is important, therefore, that holistic healing be given.

Sub-Saharan traditional healing aims at restoration of balance in both natural and supernatural relations. The supernatural aspect was often done by sacrifices and specified rituals.[118] The objective for traditional healing was beyond mere treatment of the specific disease. Kasanene affirms that traditional healing practice had as its goal the "personal integration, environmental equilibrium, social harmony and harmony between the individual and both the environment and the community."[119] The healer has to ascertain the healing is comprehensive. Alongside prescribing some herbs he "has to go beyond the mere physiological and individual symptoms, until the proper psychological, moral and socially-conditioned cause can be traced and discovered."[120]

Benezet Bujo describes the practice and objective of African traditional healing as follows: "It treats disease not only with powerful medicines, but also with rituals that place the patient at the center of a social drama in which emotions are highly charged and symbolically expressed."[121] The significance of the ritual and symbols is to effect holistic, psychological, social and physical healing."[122] Since disease is perceived as a breakage of wholeness and integration within oneself with the society and the cosmos, Bujo notes on the importance of making the patient feel needed by society and the cosmos. He writes, "The afflicted person is made to feel important and the object of social concern, while the ritual also relates what is happening to her wider cosmological and social concerns."[123]

According to Bujo this African healing techniques "enhance positively the patient's psychological state—thus providing a more favorable climate for physical and psychological healing to take place."[124] The patient is affirmed as an important needed member of the society whose dignity can never be compromised. Emphasizing this point, Bujo compares African perspective on the sick person to the modern western one. He writes, "The patient is not rejected as deviant, as a malingerer or as

---

[116] Bujo (1998, p. 123).
[117] Bujo (1998, p. 182).
[118] du Toit (1980, p. 23).
[119] Peter Kasenene (1994, p. 142).
[120] Bujo (2001, p. 97).
[121] Bujo (2001, p. 123).
[122] Bujo (2001, p. 123).
[123] Bujo (2001, p. 123).
[124] Bujo (2001, p. 123).

a marginal character, as is often the case in western medicine, but is integrated fully into the continuing concerns of the community."[125]

Health care in Ubuntu culture is therefore not only comprehensive and holistic, it is always communitarian. Mbombo notes that whenever a member of the society was ill a representative group accompanies the sick person to the village of the medicine man or woman "to listen for this person, or listen with this person. When they come out of the consulting room, what the doctor has said is also the concern of those who are waiting."[126] The companionship that the group gives the sick person is expressive of the bond that each person has with the rest of the society and it is in itself therapeutic.

Although the healing process is communitarian, involving the whole community, no treatment of any two persons is alike, even if they have the same complaint. Treatment is as unique as each person's personality is unique. This can be explained by the way each person relates with other persons, nature, the cosmos and spirits—including ancestral spirits is different. Disease or an illness is a mere symptom of an underlying cause, which is generally a breakage of relationship and its consequent disharmony. Since relationships are never identical, illnesses may appear similar but healing process is conditioned by the cause of the disharmony, agents involved, its nature and its extent.[127]

It is not surprising that Pal states that African perspective on healing is based on a framework that is "seemingly antithetical to a quantitative biomedical framework."[128] African traditional healing is based on their perspective on reality as a whole. Senghor represents a similar perspective when he states that Africans refuse to draw a line between themselves as subjects and their object of reason or act. They would rather see the threads of interconnectedness between all that exists.[129] A major part of healing is reconciliation which goes beyond the visible community and reality to the invisible, among the living and between the living and the dead.[130] Rituals for healing continue even when the patient dies. Austine Okwu enumerates some of the rituals that go beyond burial. They include: "charity cooking and eating, exchanging good wishes, confessions, the patrilineal head's blessing, and, in some cases societal singing and dancing, are forms of psychotherapeutic drama."[131]

1.3.3 Ubuntu Communitarian Healthcare Ethics

In Ubuntu culture, the sick and the people with disabilities are always a responsibility of the society. The proximate society of the sick person takes charge and accom-

---

[125] Bujo (2001, p. 123).

[126] Mbombo (1996, p. 115).

[127] Pal (2002, p. 519).

[128] Pal (2002, p. 519).

[129] See Murove (2005, p. 153).

[130] Bujo (2001, p. 97).

[131] Okwu A. S. O. 1979. Life, death, reincarnation, and traditional healing in Africa. *A Journal of Opinion* 9 (Autumn, 1979) 19–23.

panies their sick until he/she gets well or dies. Human company, its empathic presence, and solidarity with the sick or dying member of the society manifest each participant's and the community's moral maturity. If the patient is dying, the caregiver (which is the entire immediate community) helps him go through their initiation and incorporation into the community of the living dead. They are thus "included in the *process of personal growth* even as their physical strength declines."[132] Usually the objective of the community is to provide the sick and the dying with courage, peace and dignity while easing their physical, emotional or psychological pain as much as possible.[133] The practice of accompanying the sick and the dying is not only considered as virtue or charity, it is an obligation, a responsibility and duty of all members of the society. For the sick person the supportive presence of other members of the community is "a manifestation of a unified concept of the individual, in which he or she is not isolated, but part of others."[134]

Caring for the sick, the aged and the dying is considered to be a practical learning experience. Since there is no formal school in the traditional society, Alasdair MacIntyre relates that "education in virtue and the promotion of ethical living are tasks incumbent upon the entire community, and this implies that the community gives expression to itself through each individual action."[135] There are other informal ways of learning such as riddles, proverbs and myths. Parrinder describes one of them, Myth. "Myths are stories, the product of a fertile imagination, sometimes simple, often containing profound truths… some of these are philosophical, in that they consider great questions such as the meaning of life, the origins of all things, the purpose and end of life, death and its conquest.

These are often the subject of myths, which are philosophy in parables."[136] Indigenous sub-Saharan learning was not considered successful until it was put into practice. Caring for the needy, in this case the sick, elderly and dying occasioned a moment of testing how the community as a whole and its individuals are well trained in life's important facts. By caring for the sick, the old and the dying, the community members learn not only what happens to human beings but, above all, they learn the joy of giving; that is living for others and through others.

On the part of caregivers, the companionship provided to the sick and the dying is an act of worship to God. There is no separation between these different aspects of the same act. Kasanene notes that there is really no separation between religion and ethics, between one's beliefs and one's actions towards others. Ethics is an integral part of religion.[137] Since God is considered both transcendent and immanent, it was not permissible to separate the secular from the religious, matter and spirit.

---

[132] Mbombo (1996, p. 114).

[133] Mbombo (1996, p. 114).

[134] Berg (2003, p. 200).

[135] MacIntyre, A. 1985.*After virtue: A study in moral theory,* 205–206. Notre Dame: Duckworth.

[136] Parrinder (1967, pp. 15–16).

[137] Kasanene (1994, p. 140).

Caring for the dying gives worship to God as praying or offering libation through ancestors also gives worship to God.[138]

The ideal of ethical maturity is not self-centered but other-centered. Tutu describes a mature person as "open and available to others, affirming of others, does not feel threatened that others are able and good, for he or she has a proper self-assurance that comes from knowing that he or she belongs in a greater whole and is diminished when others are humiliated or diminished, when others are tortured or oppressed."[139] Consequently, Tutu describes Ubuntu as "the essence of being human" since it "speaks of the fact that my humanity is caught up and is inextricably bound up in yours. 'I am human because I belong.'"[140]

1.4 Conclusion

This chapter has briefly explored the essence of Ubuntu in view of demonstrating its usefulness to Global Bioethics. Ubuntu represents many indigenous African cultures that can make original contributions to Global Bioethics. Ubuntu provides insights about human relationships with communities and the world that enable this indigenous African ethics to contribute to Global Bioethics.

Ubuntu's fundamental objective is life itself. The unity that Ubuntu advocates is geared towards support of the dignity of human life. As such, harmonious community living which enables and promotes each life is a moral and religious obligation. Human dignity demands human rights. Human rights are based on the rights of the biosphere, since human beings cannot be extricated from the biosphere. Individuality and personhood are facilitated by context, which comprises human society, biosphere, and the cosmos. Maintaining the integrity of the biosphere and the cosmos cannot be overemphasized. Biosphere and the cosmos interrelate in a way that ultimately supports human life. Community is of utmost importance because of its role in promoting and supporting each life in it. The ultimate personal moral obligation is to become fully human, which in Ubuntu, means entering increasingly into community with others without losing one's individuality. Cut off from human community, personal consciousness development and actualization is impossible. Maintenance of harmonious optimal symbiotic interrelationships between humans, the biosphere and the cosmos is the ideal of Ubuntu.

There is a sharp contrast between Ubuntu worldview and Cartesian ontology and epistemology. From the perspective of Ubuntu, all knowledge, including self-knowledge, is other-oriented. The self has to be objectified to be accessed, even as it remains the subject. Relationship is inescapable in any real human existence. All ethics and morality are based on accepting otherness and relating to it. The relationship between the self and the inescapable other can be morally evaluated.

---

[138] Mbiti (1970, p. 1).

[139] Tutu (1999), cited by Battle (2009, p. 2).

[140] Tutu, Desmond. http://www.tsabcc.org/ubuntu/philosophy.htm.

Consequently, devoid of all relationship, there is no human life. This explains why Ubuntu treasures interdependence, initiations into wider circles of the society and the cosmos. Such relationships go on beyond physical death into the world of the *living-dead* and eventually spirits and divinities. To live, therefore, is to relate. Morality evaluates and explains human relationship with self, other humans, the biosphere and the cosmos.

The centrality of heterosexual marriage in Ubuntu has been explored. Human race is in duality of man and woman, which duality relate intimately to generate more life. Heterosexual marriage is not a mere option but a moral obligation for the survival of human species. It is unethical not to marry and generate progeny. Since humans receive their existence from their predecessors, they have an obligation to be a link in the chain that assures survival of the human species by producing progeny. Giving back to the community that way is expected of everyone. Hence, marriage and sexuality are not necessarily private affair. They are a concern of the entire community.

Responsible human relationship is at the core of Ubuntu, enabling Ubuntu to be comparable to Care Ethics. The sick, those with disabilities and the poor are a concern of everybody. Justice is secondary to, and part of, care. Ubuntu understands human disease comprehensively, essentially as a breach or breakage of human integrity. Ubuntu healthcare addresses not only the visible symptoms, but the possible underlying physiogenic, psychological, social and ontological causes. Healing is a process of reconciliation. Healing reconciles and restores the lost unity within the self, between the self and the society, between the self and the diseased, between the self and the cosmos and between the self and God. Ubuntu perspective on human disease and healing is comprehensive and holistic.

Chapter 1 has demonstrated that healthcare in Ubuntu is a concern of all members of the society. Caring for the sick is not charity. It is an ethical obligation. It is a proof of one's moral maturity. For the sick person, the empathic and supportive presence of the community confers a feeling of belonging and sharing in the life of the whole community, even as their individual life declines. In the event that the sick person is terminal, their sickness is approached as a process of initiation into the world of the *living-dead*. Their decline becomes an eschatological hope-filled process of personal growth into the destiny of human life. The role of the community is to give dignity and courage to the dying and preparing the living to face their own mortality as they help others through the process. The following chapter will study Ubuntu more deeply by exploring its three constitutive components in view of demonstrating better how Ubuntu can contribute to, and be enlightened by, Global Bioethics.

Chapter 2
Ubuntu Ethics

Beauchamp and Childress define the term ethics as a "generic term covering several different ways of examining and understanding the moral life."[1] Childress and Macquarrie describe ethics and ethical questions in three different ways. The first are "questions as to what is right, good, etc, or of how we ought to behave (normative ethics, morals)." The second are "questions as to the answers given by particular societies and people as to what is right or good." The third are "questions as to the meanings or uses of the words used in answering questions of what is right, good."[2] Emmet describes morality as "Considerations as to what one thinks it important to do and in what ways; how to conduct one's relations with other people; and being aware and prepared to be critical of one's basic approvals as disapprovals."[3] Dewey asserts that "interest in learning from all the contacts of life is the essential moral interest."[4]

As an ethic, Ubuntu is generally in conformity with the definitions and descriptions of ethics given above. Ubuntu, however, is unique in its substance, in its method and in its worldview. As an indigenous culture Ubuntu presents an ethical worldview (referred to in this work as Ubuntu ethics) with three constituent components. The first component of Ubuntu ethics deals with the tension between individual and universal rights; the contribution of this component to global bioethics emerges by considering the Ethics of Care as a crucial aspect of bioethics discourse.

The second component of Ubuntu ethics concerns the cosmic and global context of life; the contribution of this component to global bioethics emerges by considering UNESCO's Universal Declaration on Bioethics and Human Rights as crucial for bioethics discourse. The third component of Ubuntu ethics deals with the role of solidarity that unites individuals and communities within a cosmic context; the contribution of this component to global bioethics emerges by considering the Roman Catholic tradition on social ethics as a significant aspect of discourse on global bioethics. This chapter explores those three major components of Ubuntu ethics.

---

[1] Beauchamp and Childress (2009, p. 1).

[2] Childress and Macquarrie (1986, p. 206).

[3] Emmet (1979, p. 7).

[4] Dewey (1929, p. 418).

L. T. Chuwa, *African Indigenous Ethics in Global Bioethics,* Advancing Global Bioethics, 33
DOI 10.1007/978-94-017-8625-6_2, © Springer Science+Business Media Dordrecht 2014

2.1   Tension Between Individual and Universal Rights

The first major component of Ubuntu concerns the tension between individual and universal rights. The meaning of this context is enlightened by considering the Ethics of Care. This component has three related concepts. The first concept is inalienable rights. Every human individual has inherent inalienable rights to be recognized and respected by other human beings. The second component is human relationships. Recognition of personhood necessitates the development of human relationships with other persons in the society and with the society as a whole. The third concept is reciprocity of care. Fostering reciprocity of care occurs through personal acceptance and assumption of duties and responsibility in society.

2.1.1   Inalienable Rights

Ubuntu protects the inalienable rights of individuals. Each person's uniqueness is connected with rights and obligations.[5] However, individual rights are only recognizable in the context of society.[6] In Ubuntu culture every human being is entitled to all basic human rights. However, there is a very deep implied understanding that personal human rights are subordinate to, and dependent on, the basic communitarian interests and wellbeing.[7] Even if a person has inalienable rights such as right to life and to personal human dignity, it is the community that recognizes those rights. There is, therefore, a tension between individual human rights and societal basic rights and interests.

2.1.1.1   Personal Rights within Communitarian Context

One of the greatest scholars of African communitarianism is Leopold Senghor from Senegal. In his view Africans view community as precedent to its component individuals. Consequently the community is more important than it's the individuals who make it. Likewise, according to Senghor's views, solidarity should take precedence to individual decision and activity. Community needs should be precedent to individual needs. He contends that Africans place more emphasis on the "communion of persons than on their autonomy."[8] In his work titled *Consciencism*, Nkrumah argues that from the African perspective everything that exists is in a complex web of dynamic forces in tension but with necessary interconnection and complementarity.[9] Nkrumah's views are consistent with Senghor's observation of

---

[5] Macquarrie (1972, p. 110); Shutte (1993, p. 49, 51).

[6] Holdstock (2000, pp. 162–181).

[7] Asante et al. (2008, p. 115).

[8] Senghor (1964, p. 49, 93–94).

[9] Hord and Scott Lee (1995, p. 58).

the African worldview. However, Nkrumah emphasizes the inevitable conflict and tension within the African ideal of universal unity in Ubuntu culture while Senghor places greater emphasis on the importance of societal and cosmic unity within African culture.[10] Both authors shed light on the examination of the conflict between individual and universal rights while simultaneously considering the individual's inalienable rights.

Gyekye explores the tension between basic personal rights (autonomy, freedom and dignity) and the underlying need for the society in realization of individual's potential.[11] Gyekye states that there is a relationship between the individual and the society which is reflected in the "conceptions of social structure evolved by a community of people."[12] To explain the relationship between the society and the individual, Gyekye cites an Akan proverb which goes, "The clan is like a cluster of trees which, when seen from afar, appear huddled together, but which would be seen to stand *individually* when closely approached."[13] This proverb is an analogy which implies that even though some branches of the trees may touch, or even interlock each tree stands individually and has its own identity. Relationships in Ubuntu should not overshadow the importance of individual autonomy. There is need for discernment and distinction of the delicate balance between the two aspects of Ubuntu.

In sum, Gyekye observes an inevitable symbiotic mutuality between personal inalienable rights and the society. The society is a needed context for realization of personhood and self-actualization. However, "Individuality is not obliterated by membership in a human community."[14] Each individual retains his or her uniqueness and basic human rights regardless the role and importance of community to the individual. According to Gyekye "the most satisfactory way to recognize the claims of both communality and individuality is to ascribe to them the status of an equal moral standing."[15]

The Ubuntu ideal of maturity is such that one retains one's individual rights without losing touch with the community which facilitates individuality. Ntibagirirwa states that Ubuntu arms one with "normative principles for responsible decision-making and action, for oneself and for the good of the whole community."[16] Individualistic action which leaves out the community would consequently be unethical. Once an individual has acquired enough ethical maturity to act simultaneously for self and for the community, such person is considered morally mature. In the words of Ntibagirirwa, "S/he no more does things because the community expects him/her to do so, but because it is the right thing to do for both him/herself

---

[10] Hord and Lee (1995, pp. 46–50).

[11] Gyekye (1997, p. 35).

[12] Gykye (1997, p. 35).

[13] Gykye (1997, p. 40).

[14] Gykye (1997, p. 40).

[15] Gykye (1997, p. 41).

[16] Ntibagirirwa (1999, p. 104).

and the community."[17] In Ntibagirirwa's view "It is Ubuntu alone that can allow the individual to transcend, when necessary, what the customs of the family or the tribe requires without disrupting the harmony and the cohesion of the community."[18]

2.1.1.2 Individual's Personal Rights are Defined by Others' Personal Rights

One of the criticisms against Ubuntu is that it limits personal autonomy and freedom. On the contrary, Ubuntu champions realistic ethical freedom. Weil explains this position when he states that "It is not true that freedom of one man is limited by that of other men." Freedom is always relative to the freedom of others. "Man is really free to the extent that his freedom fully acknowledged and mirrored by the free consent of his fellow men finds confirmation and expansion of liberty. Man is free only among equally free men." Ubuntu recognizes the fact that "the slavery of even one human being violates humanity and negates the freedom of all."[19] Freedom in particular and virtue in general, therefore, are contingent to, and defined by community society and the common good. No individual is greater than the society; individual members of the society are parts of, and enabled by the society. However, Kasanene notes, "individuals are able to think and act independently, as long as their actions do not harm others, and so the individual has to always bear in mind that excessive individualism is regarded as being a denial of one's corporate existence."[20]

Thus, strictly speaking, from the perspective of Ubuntu there can be no absolute individual rights. All individual rights are understood within the matrix of the community. Consequently, Kamwangamalu argues that Ubuntu is communitarian since "the group constitutes the focus of the activities of the individual members of the society at large...the good of all determines the good of each or... the welfare of each is dependent on the welfare of all."[21] Since the individual rights are based on, and facilitated by, common good, individuals in the culture of Ubuntu should act for themselves and the community rather than for themselves against the community. The tension between individual rights and the community is resolved by considering inalienable individual rights in the context of societal common good.

2.1.2 Human Relationships

Ubuntu protects human relationships. Although personhood is intrinsic and innate to human beings its recognition is of vital importance. Morality is based on mutual recognition of personhood in any human parties in relationship with each other.

---

[17] Ntibagirirwa (1999, pp. 104–105).

[18] Ntibagirirwa (1999, p. 104).

[19] Weil (1973, p. 182, 188–189).

[20] Kasanene (1994, p. 143).

[21] Kamwangamalu (2008, p. 115).

Thus, independent of human relationship the innate personhood in human beings remains only potential.[22] In Ubuntu culture, it is the community that defines a person by judging whether one has attained full moral maturity. This judgment is based on the individual's relationships with the community, that is, whether one has moral values, feelings and empathy that facilitate others' wellbeing. One contributes to the definition of oneself through everything one does. A person's identity or social status and the rights that are attached to that identity go hand in hand with that person's responsibility or sense of duty towards, and in relation to, others.[23]

2.1.2.1 Anthropological and Epistemological Perspective

In order to understand Ubuntu ethics, one has to first understand African anthropology and epistemology. One of the most important clues into Ubuntu mindset is an insight into the African traditional way of thinking. Traditional African thinking is "not in 'either/or,' but rather in 'both/and' categories."[24] The second clue is related to the first. That is, understanding the primacy of community in Ubuntu ethics. Bujo recognizes "community as a starting point in African ethics."[25] John Macquarrie explains that in Ubuntu individuals can only exist as human beings in their relationship with other humans. The word "individual" therefore, "signifies a plurality of personalities corresponding to the multiplicity of relationships in which the individual in question stands." Hence, "being an individual by definition means 'being-with-others.'"[26] The phrase 'being-with-others' in itself defines the nature of the relationship either as good or bad, right or wrong. It is evaluative. Relationships reveal how beneficent the parties are.

2.1.2.2 Otherness

To underline the importance of human relationship in the culture of Ubuntu, Van Der Merwe emphasizes the importance of the concept of otherness, which implies relationship. He observes that the African worldview is based on the understanding that "A human being is a human being through the *otherness* of other human beings."[27] This observation is far reaching in Ubuntu Ethics since it is the 'otherness' of another human which helps to prove one's humanity. Consequently, personal maturity is measured by the way one relates with others. That is, self-actualization happens in the process of fulfilling one's obligations and duties toward others. Menkiti states that assumption of responsibility towards others "transforms one from the *it*-status of early child-hood, marked by an absence of moral function, into the personhood

---

[22] Shutte (1995, p. 46); Holdstock (2000, pp. 162–181).

[23] Mnyaka and Motlhabi (2003, p. 224).

[24] Bujo (2001, p. 1).

[25] Bujo (2001, p. 1).

[26] Macquarrie (1972, p. 104).

[27] Van Der Marwe and Willie (1996, pp. 1–3).

status of later years marked by a widened maturity of ethical sense—an ethical maturity without which personhood is conceived as eluding one."[28]

Due to the importance of "otherness" in self-recognition, self-actualization and moral development, human relationship is vital in the culture of Ubuntu. It is the community which defines a person and enables that person to find the self through the vehicle of human relationships. Thus, there is a delicate balance between individual autonomy and the role of society in personal life within Ubuntu culture. Using the words of Macquarrie, true Ubuntu "preserves the other in his otherness, in his uniqueness, without letting him slip into the distance."[29] This statement indicates the role and importance of human mutuality and interdependence. The self always stands in need of an-other both for the self and for the other, since there cannot be self without an-other.

2.1.2.3   Communitarianism

One of the distinguishing features of Ubuntu ethics is the significant role of community in comparison to that of individuals in any particular ethical situation. Ubuntu ethics is based on, has as its goal, and is validated by societal common good. The role of community in Ubuntu ethics is based on the premise that none of community members would be what he or she is without the community. Thus, naturally the community takes precedence over the individual without underestimating individual personal rights. Teffo argues that Ubuntu "merely discourages the view that the individual should take precedence over the community."[30] The objective of Ubuntu ethics is the balance between individual rights and the necessary communitarian conditions which facilitate and support those rights.

Each member of the community has a right to self-determination which finds its limitation in common good. The justification of this assertion is given by a number of Ubuntu scholars. Michael Battle argues that personhood happens through other persons. He observes that "we don't come fully formed into the world…we need other human beings in order to be human. We are made for togetherness; we are made for family, for fellowship, to exist in a tender network of interdependence."[31] Mkhize states that "the African view of personhood denies that a person can be described solely in terms of the physical and psychological properties. It is with reference to the community that a person is defined."[32] However, Ubuntu neither overlooks nor underestimates individual self-determination.

Macquarrie, writing in *Existentialism,* cautions against a misunderstanding of Ubuntu. He states that when communitarianism becomes oppressive, then Ubuntu is

---

[28] Menkiti (1984, p. 172).

[29] Macquarrie (1972, p. 110); Shutte (1993, p. 49, 51).

[30] Teffo (1994, p. 7, 12).

[31] Battle (1997, p. 65).

[32] Nhlanhla Mkhize, "Culture, Morality and Self, In Search of an Africentric Voice," Cited in, Barbara (2003) http://www.barbaranussbaum.com/downloads/reflections.pdf, February 15, 2012.

abused. Ubuntu respects individual autonomy, "true Ubuntu incorporates dialogue. It incorporates both relation and distance." Ndaba addresses the two aspects of Ubuntu when he argues "that the collective consciousness evident in the African culture does not mean that the African subject wallows in a formless, shapeless or rudimentary collectivity…it simply means that the African subjectivity develops and thrives in a relational setting provided by ongoing contact and interaction with others."[33]

Because of the role of community and human relationships in Ubuntu, Nkonko Ka-mwangamalu argued that Ubuntu is communitarian since, in his view, the society dictates "not only the rights of an individual but also individual's duties, obligations and limitations/boundaries."[34] What underlies this observation, however, is the important role of human relationship in Ubuntu culture. In his work, *Ubuntu in Comparison with Western Philosophies*, Ndaba asserts that "African subjectivity develops and thrives in a relational setting provided by ongoing contact and interaction with others."[35] Ndaba's assertion, however, is not limited to Africans. All human beings stand in need of human interaction for their personal actualization and thriving of the society.

2.1.3 Reciprocity of Care

Ubuntu fosters reciprocity of care. Individual/universal human rights are conjoined with human reciprocity of care and the assumption of responsibility.[36] All beings exist in reciprocal relationship with one another. In Ubuntu culture every individual has an irreplaceable role to play. Everything that exists contributes to the equilibrium necessary for sustenance of ecosystems and integrity of the biosphere and the cosmos.[37] It is the reciprocation which facilitates individual, societal and the biospheric survival and progress. Proper reciprocation generates harmony while failure to do so may generate violence.[38] Reciprocity is a sacred duty. Exploitation is unethical and immoral. Life from this perspective is only real if it is shared and shares in the lives of others. In his work *Ubuntu Management and Motivation*, John Broodryk notes that Ubuntu is both a state of being and of becoming, both of which are anchored in reciprocity of care, thus as a process of self-realization through others, Ubuntu enhances the self-realization of others.[39] Ethics of Ubuntu rest on the assumption that as one is enabled by the community to find oneself and grow as human person, one should use one's potential for the good of the community. Life is about receiving and giving. Failure to reciprocate is tantamount to violence. It is unethical.

---

[33] Teffo (1994, p. 7, 12).

[34] Kamwangamalu (2008, p. 115).

[35] Ndaba (1994, p. 14).

[36] Van Der Marwe and Willie (1996, pp. 1–3).

[37] Richards (1980, pp. 76–77).

[38] Richards (1980, pp. 76–77).

[39] Broodryk (1997, pp. 5–7).

2.1.3.1   Reciprocity as the Bond Between the Community and an Individual

Broodryk posits that, "as a process of self-realization *through* others, Ubuntu enhances the self-realization *of* others."[40] Macquarrie observes that "being with others…is not added on to a pre-existent and self-sufficient being; rather, both this being (the self) and the others find themselves in a whole wherein they are already related. By nature, a person is interdependent with other people. Due to this interdependence, reciprocity is *sine qua non* within the culture of Ubuntu. By nature a person receives and reciprocates care. The community or society is a prerequisite for personhood. Society facilitates reciprocation which, in turn, facilitates personhood and self-actualization. Personal reciprocation of care creates, sustains and strengthens the community. Reciprocity in form of giving back to the community and proactive living for the community and others defines a person and his moral maturity. This approach to morality is unique since it defines personhood for community not *against* community. Macquarrie explains this perspective in detail in his work titled *Existentialism*.[41]

Morality is about human relationships while a human relationship is about reciprocity. Wrong doing separates people, disturbs harmony, and is against life. Verhoef and Michel, in their article titled "Studying morality within the African context," assert that "what is right is what connects people together; what separates people is wrong."[42] Now what connects people together involves reciprocity since human relationship is anchored on reciprocity. In agreement with Verhoef and Michel, Thaddeus Metz identified a concise ethical principle based on African relationality, solidarity and reciprocity: "an act is right just insofar as it is a way of living harmoniously or prizing communal relationships, ones in which people identify with each other and exhibit solidarity with one another; otherwise, an act is wrong." In other words indigenous sub-Saharan ethics' (Ubuntu) objective is harmony which favors human life. Harmony, however, is a product of mutually favorable human actions. Reciprocity is a necessary component in sub-Saharan ethics. Metz explains solidarity with one another as "to act in ways that are expected to benefit each other…solidarity is also a matter of people's attitudes such as emotions and motives being positively oriented toward others, say by sympathizing with them and helping them for their sake."[43]

2.1.3.2   Ujamaa as Praxis of Ubuntu Reciprocity

Many post-colonial African intellectuals tried to force Ubuntu into a political theory. Politicians such as Julius Nyerere[44] of Tanzania, Kwame Nkrumah[45] of Ghana

---

[40] Broodryk (1997, pp. 5–7).

[41] Macquarrie (1972, p. 104).

[42] Verhoef and Claudine (1997, p. 397).

[43] Metz (2010, p. 84). http://www.tandfonline.com/loi/cjhr20, February 15, 2012.

[44] See Nyerere (1968, 1973); Russian Academy of Sciences Institute for African Studies (2005).

[45] See Nkrumah (1964). Although Nkrumah's objective was to help Africa deal with the changes from Islam and the West without losing its Identity, Ubuntu remains the most important element within African cultural identity.

and Leopold Senghor[46] of Senegal are some of the leading examples. Their zeal for Ubuntu as a political theory failed to come to fruition primarily because Ubuntu, being an ethic, could not be reduced to a political ideology. This section explores Nyerere's *Ujamaa*, a Swahili word for familyhood or fraternity, (which Nyerere interpreted as African socialism) as praxis of Ubuntu reciprocity.

In Nyerere's own words, Ujamaa "is an attitude of mind." It is that "attitude of mind, and not the rigid adherence to a standard political pattern, which is needed to ensure that the people care for each other's welfare."[47] Ujamaa is about care and reciprocity. Nyerere, while trying to show that Ujamaa is socialism, ended up demonstrating that it really is not. Contrasting socialism and capitalism to justify Ujamaa as socialism Nyerere writes: "Destitute people can be potential capitalists—exploiters of their fellow human beings. A millionaire can equally well be a socialist; he may value his wealth only because it can be used in the service of his fellow men." This statement of Nyerere not only contradicts the meaning of socialism, it affirms Ujamaa as Ubuntu ethic. While socialism is imposed on the people, Ubuntu is a cultural ethic, not a political ideology. Nyerere describes such ethic. He paradoxically further describes it even as he contrasts socialism from capitalism. Nyerere writes, "The man who uses wealth for the purpose of dominating any of his fellows is a capitalist. So is the man who would if he could. ...a millionaire can be a good socialist."[48] Nyerere argued that Ujamaa "is opposed to capitalism, which seeks to build a happy society on the basis of the exploitation of man by man; and it is equally opposed to doctrinaire socialism which seeks to build its happy society on a philosophy of inevitable conflict between man and man."[49] What Nyerere neither defines nor explains in detail is the meaning of Ujamaa. By his own statements with regards to Socialism and Capitalism, Nyerere shows that Ujamaa is an attitude of mind and a moral mindset. It is not a socio-political and economic theory. Ujamaa is an ethic. As an ethic, Ujamaa transcends political and economic theories and systems. Ujamaa is simply praxis of Ubuntu. It is essentially an ethic.

In the traditional society, everybody who was able to work had to work hard for personal needs and the needs of the sick, the old and children. Provision for those who could not provide for themselves was imperative. The traditional society didn't force its constituents to distribute their produce. It did not emphasize equality of possession but of personhood. Recognition of human dignity and personhood in all humans, including those with disabilities, and safeguarding that dignity is the ethical ideal of both Ujamaa and Ubuntu. Thus, individual ownership of major means production such as land was discouraged but without the use of force or

---

[46] See Washington (1973). Senghor uses the concept of Negritude in poetry to explore African culture, the basis of which is Ubuntu. Some of his main concepts include human and cosmic unity, rhythm, importance of human emotion and the power of art to communicate what cannot be easily verbalized.

[47] Nyerere (1968, p. 1).

[48] Nyerere (1968, p. 1).

[49] Nyerere (1968, p. 12).

political ideology.[50] People were allowed to participate in the process of production of wealth according to their ability. Consequently, there was naturally a division of labor and subsidiary.

Traditional Ujamaa gave members of its respective society, specifically people with physical disabilities, the less fortunate, the old, children and the sick the security they needed to live a meaningful and dignified life in spite of their limiting conditions. Nyerere argues that such security which was common in, almost all traditional societies must be preserved and extended beyond tribal, national and continental boundaries because all people are equal.[51]

The Arusha Declaration was founded on the traditional African way of life. The declaration recognizes human equality, human right to life, dignity and respect; equal rights as citizens, equal right of expression, movement, religious belief, right of association, right to be protected by the society, right to just reward for human labor, equal right of access to national natural resources and major means of production.[52]

In sum, Ujamaa is systematized Ubuntu in praxis. Ujamaa is based on the need to recognize human equality and the ethical imperative of investing in the community based on each individual's need for the community and the community's need for its constituents. It is ultimately about giving back to the community, for the good of all, without denying personal rights and entitlements.

2.1.3.3   Importance of Marriage and Procreation

Most traditional African societies hold marriage as the focus of both individual and societal existence. Mbiti observes that in marriage all members of the society, the living, the dead and the yet to be born meet. Whoever does not participate in it "is a curse to the community, he is a rebel and a law-breaker, he is not only abnormal but 'under-human'. Failure to get married under normal circumstances means that the person concerned has rejected the society and the society rejects him in return."[53]

From the individual's perspective, the importance of marriage is based on the belief that parents are reproduced in their progeny, which means parents with children will be immortal as long as their children don't break the chain by not making children.[54] Having descendants is also crucial because one's immortality (in the world of the living-dead) is acquired by having descendants who will keep the diseased in memory. "To die without getting married and without children is to be completely cut off from the human society, to become disconnected, to become an outcast and

---

[50] Nyerere (1968, pp. 2–12).

[51] Nyerere (1968, p. 12).

[52] Nyerere (1968, p. 14). The Arusha declaration was passed on February 5, 1967. Being derived from the traditional society way of life, the Arusha declaration proves not only the inherent ethics in the traditional society but also its authenticity and validity as compared to modern ethics.

[53] Mbiti (1969, p. 130).

[54] Mbiti (1969, p. 130).

to lose all links with mankind."[55] Naturally, therefore, the society hopes and expects that everybody marries and begets children. Each person has an ethical obligation to marry both for the sake of the self and of the community.

Traditionally, the society improvised a system whereby a couple who have biological impediment such that they cannot have children of their own could have children who would keep them alive in their memory after they die. In patrilineal societies, a brother or another designated close relative of the childless deceased or incapable parent would help by having intercourse with the wife of the deceased or the incapable parent for the purpose of making children for him.[56] Bujo asserts that "the main presuppositions of African ethics are not the same as those involved in natural-law approaches. The main goal of African ethics is fundamentally life itself. The community must guarantee the promotion and protection of life by specifying or ordering ethics and morality."[57] Marriage is the main way the community fulfills its duty to promote life.

The centrality of marriage is based on the event in which two persons willingly express their desire to cooperate to keep the society immortal. Most peoples in Africa south of the Sahara hold that humans owe their existence to many generations of ancestors. There are many sayings to the effect that we received our existence from them and we must in turn give existence to the next generation. Marriage is an ethical responsibility and a religious sacred obligation. We walk on the graves of our ancestors; we should let others (our progeny) walk on our graves. We stand on their shoulders. It is their selflessness, best expressed in marriage, that they generated progeny. Marriage is the unique opportunity that reveals a couple's willingness to give back to the society by accepting the role of keeping the chain of generations going. Failure to do so contributes to killing of the society by rendering it futureless.[58]

2.2 Cosmic and Global Context

The second major component of the culture of Ubuntu concerns its sense of a cosmic/global context. The meaning of this context is enlightened by considering the UNESCO Code of Bioethics. This component has three related concepts. The first concept is restorative justice which is necessary in order to maintain lasting peace and order. The second concept is respect for diversity in order to achieve personal and societal fulfillment. The third concept is respect for and protection of the cosmos as the context which supports the biosphere and human society.

---

[55] Mbiti (1969, p. 131).

[56] Mbiti (1969, pp. 141–143).

[57] Bujo (2001, p. 2).

[58] Mbiti (1969, pp. 130–145).

2.2.1   Justice

Most indigenous African cultures that embrace Ubuntu require restorative justice[59] which is founded on human dignity and equality within human society. Its objective is restoration of peace and order.[60] In his autobiography, Nelson Mandela explains Ubuntu restorative justice. He states that the oppressor and the oppressed both need liberation since a person who takes another person's rights is a prisoner of his own hatred and prejudice. "The oppressed and the oppressor alike are robbed of their humanity."[61] Mandela's views about human freedom, which represent the Ubuntu cultural meaning of justice, are expressed in the statement, "to be free is not merely to cast off one's chains, but to live in a way that respects and enhances the freedom of others."[62]

2.2.1.1   Ubuntu Justice is Reparative Rather than Retributive

Mandela's insight is shared by most liberation fighters during the era of apartheid in South Africa. Addressing the role of Ubuntu during and immediately after apartheid in his work, *Concept of Ubuntu as a Cohesive Moral Value*, Teffo expresses the general prevailing spirit, "there is no lust for vengeance, no apocalyptic retribution."[63] On the contrary there is a yearning for justice, and for "release from poverty and oppression, but no dream of themselves (black South Africans) becoming the persecutors, of turning the tables of apartheid on white South Africans."[64] The Ubuntu ideal of justice is restorative rather than retributive or punitive. Ubuntu restorative justice is founded on the understanding that human community is analogous to an organism. If one part is hurt the whole organism hurts. Restoration of tranquility, equilibrium and order is the ethical ideal. Violence is harmful not only to its direct victim, but also to the perpetrator and the society. The objective of justice in Ubuntu is peace and community building.[65] Consequently, Maphisa attributes that the transformation of an apartheid South Africa into a democracy to what he termed "a discovery of Ubuntu."[66]

Thaddeus Metz observed an unwritten ethical principle in sub-Saharan peoples that most African communities South of Sahara hold that it is immoral "to make policy decisions in the face of dissent, as opposed to seeking consensus."[67] In case

---

[59] Mandela (1994, p. 544).

[60] Teffo (1994, p. 11).

[61] Mandela (1994, p. 544).

[62] Mandela (1994, p. 544).

[63] Teffo (1994, p. 5).

[64] Teffo (1994, p. 5).

[65] Van Der Marwe (1996, p. 1).

[66] Maphisa (1994, p. 8).

[67] Metz (2007, p. 324).

of dispute, there is no clear distinction between conflict resolution and execution of justice. The resolution process aims at mutual education, community education, character formation and consensus seeking. Since the objective is reparation and restoration of peace and harmony, the parties, along with the rest of the community, engage in active reflective listening and the discussion continues "until a compromise is found and all in the discussion agree with the outcome."[68] Dispute and conflict occasion a moment to teach and reinforce virtues of Ubuntu. Tutu describes a virtuous person from the perspective of Ubuntu as "welcoming, hospitable, warm and generous, willing to share."[69] Elsewhere he describes such a person as are "open and available to others, willing to be vulnerable, affirming of others, do not feel threatened that others are able and good, for they have a proper self-assurance that comes from knowing that they belong in a greater whole."[70] Ubuntu sense of justice is, at the same time, educative and community building.

In sum, Ubuntu justice is restorative since it is based on the maxim "I am human because I belong. ...my humanity is caught up and inextricably bound up in yours."[71] Because of such interconnection and symbiotic interdependence, virtuous persons know that "they are diminished when others are humiliated, diminished when others are oppressed, diminished when others are treated as if they were less than who they are."[72] The objective of criminal justice in Ubuntu is reconciliation, not retribution.[73] As a result, from the perspective of Ubuntu, retributive punitive justice is unethical and counterproductive. It is destructive of the ideal and objective Ubuntu.

2.2.1.2 Ubuntu Justice is Distributive

Ubuntu is radically opposed to libertarian philosophy represented by Locke regarding property and individual liberty. According to Locke, property means both material possessions and liberty.[74] The concept of property is the kernel of individual freedom. Civil government is a product of social contract whose objective is to ensure protection of private property from the encroachment of others. Lockean freedom, therefore, simply means control and possession of one's own person and possessions.[75] American tradition has historically placed great faith in the Lockean vision of the individual with its emphasis on negative freedom and private property

---

[68] Metz (2007, p. 324).

[69] Tutu (2009, p. 2).

[70] Tutu (2009, p. 2).

[71] Tutu (http://www.tsabcc.org/ubuntu/philosophy.htm. February 15, 2012).

[72] Tutu (2009, p. 2).

[73] Metz (2007, p. 325).

[74] J. Tully, *A Discourse on Property: John Locke and His Adversaries* (Cambridge: Cambridge University Press, 1980), pp. 38–50, 59–63.

[75] A. M. C. Waterman, "Property Rights in John Locke and in Christian Social Teaching," *Review of Social Economy* 11, no. 2 (1982): pp. 97–115.

rights.[76] Nozick agrees with Locke in many ways. In his view, distributive schemes unjustly redistribute assets already owned by individuals, without taking into account the way in which assets have been acquired.[77] Most tax redistributions to fund health care or any other need are unfair. They fail to recognize that individuals are entitled to their personal holdings.[78] This position implies that the poor may be unfortunate but their plight is not a moral problem. They have no just claim to others' entitlement.[79] Due to the Lockean influence in American thought, the legacy of the firm entrenchment of property rights led to an exaggerated importance of the concept of individual property rights over the claims of other human values such as equality and fraternity.[80] The healthcare insurance market can be characterized in the very same terms.[81]

Rawls sought to resolve conflicts between the values of liberty and equality based on fairness. He argued for an original position in which individuals were considered to be under a veil of ignorance such that they were ignorant as their specific interests.[82] The individual in this original state is free, rational and essentially self-interested. The aim of this imaginary original position is to question what would the individuals under the veil choose as a principle for guiding justice. In Rawls' view, two principles would emerge: first, each person would have the most extensive liberty compatible with similar liberty for others. Second, social and economic inequalities would be ordered so that they are to everyone's advantage and be attached to positions open to all.[83]

Daniels further develops Rawls' concept of justice as fairness in the context of health care provision.[84] Daniels emphasizes an equality of opportunity range and the need for a basic level of normal species functioning to provide for the degree of equality of opportunity. Health care that promotes the normal range of species functioning can be justified for all on the basis of a commitment to the idea of equality of opportunity.[85] Daniel's views are in conformity with the Ubuntu ethics.

Ubuntu is more agreeable to welfare liberalism and Rawls concept of justice. Welfare liberalism challenges the classical liberalism of Locke. It is represented by Charles Fried,[86] Allen Buchanan,[87] Norman Daniels,[88] and the President's

---

[76] Mary Ann Glendon, "'Absolute' Rights: Property and Privacy," *The Essential Communitarian Rader* (Lanham, Md: Rowman and Littlefield, 1998), pp. 107–114.

[77] Nozick (1974, pp. 118–163).

[78] Nozick (1974, pp. 167–169).

[79] Nozick (1974, pp. 233–235).

[80] Ryan (1976, pp. 126–141).

[81] O'Keeffe (1994, pp. 35–64).

[82] Rawls (1971, pp. 10–17).

[83] Rawls (1971, pp. 17–65).

[84] Daniels (1985, pp. 42–49).

[85] Daniels (1985, pp. 36–42).

[86] Fried (1976, pp. 29–34).

[87] Buchanan (1984, pp. 55–78).

[88] Daniels (1985, pp. 79–333).

Commission,[89] all of whom have argued for the need to ameliorate the conditions of the market and provide enablement opportunities for all. They have argued for a two-tier system as a safety net for the poor, often expressed as a decent minimum. Ubuntu goes much deeper than mere ethics of market economy.

According to Ubuntu ethics one's personhood is a potential that is realized to the degree one participates and contributes to the life of the community. Arguing for Ubuntu development theory of personhood Menkiti states that personhood is progressively realized through personal relationships and functioning in society. From his observation personhood is attained especially by doing one's obligations in the society.

In Menkiti's own words, executing one's obligations "transforms one from the *it*-status of early child-hood, marked by an absence of moral function, into the personhood status of later years marked by a widened maturity of ethical sense—an ethical maturity without which personhood is conceived as eluding one."[90] Thus, every member of the community should be an active player in the life of the community for the sake of every other person, especially those with disabilities. It is through being an active player in the life of the community that personhood is realized. Broodryk articulates that greatest personal moral obligation in Ubuntu "is to become more fully human which implies entering more and more deeply into community with others."[91]

Ubuntu community is experienced practically in sharing of the necessities that sustain human life. Gyekye notes that according to Ubuntu ethics to be a member of the community also means to be entitled to the decent minimum of means of production and property such as land and cattle.[92] Possession of property is never absolutely personal. Bujo articulates that "the final aim is never personal enrichment. Property belongs to the individual, but only so that, in case of need, it may be placed at the disposal of the community. Attached to all property is the notion of stewardship and ministry."[93]

There is no absolute right to ownership of property. For instance, one cannot spoil food that belongs to him or her. One should keep it for any person who may need it. Bujo notes that helping the needy in the traditional society is an ethical obligation. He notices the western influence and its impact on the African values. Bujo states, "Africa is of course changing under the impact of foreign cultures, but in traditional times no one questioned the obligation of clan-members to help each other, and no one was allowed to go without the necessaries of life."[94] Equally utilization of personal potential by each person through hard work was a moral obligation.

---

[89] *Securing Access to Health Care: A Report on the Ethical Implications of Differences in the Availability of Health Services* (Washington, D. C.: President's Commission for the Study of Ethical Problems in Medicine and Biomedical and Behavioral Research, 1983).

[90] Menkiti (1984, p. 172).

[91] Broodryk (1997, p. 101).

[92] Gyekye (1997, pp. 146–152).

[93] Bujo (1992, pp. 35–36).

[94] Bujo (1992, pp. 35–36).

Consequently, "any kind of laziness or parasitism was vigorously denounced." As for theft, this was never tolerated.[95]

People with disabilities, the sick, the orphaned, widows or elderly members of the African traditional society south of Sahara are naturally protected so that they don't feel insecure or inferior to the rest of the members of the society. If a member of an ethnic group is prosperous, the whole ethnic group is prosperous. If the ethnic group is prosperous each member considers himself/herself prosperous. Land is communally owned in that; no one has absolute right to it. Members of the community use it according to need. Laziness or refusal to work is a curse and source of shame to the respective individual and his/her family.[96] Although Ubuntu is not socialism, in the sense that it does not enforce equal distribution of wealth, it does not tolerate disproportionate economic inequality. The gap between the poorest and the richest is minimized for the sake of maintenance of harmony in the community.[97]

Creation of wealth is a duty that all have to fulfill. However, there is always division of labor so that the principle of subsidiarity is naturally in operation. Everybody participates in the community in what he/she does best. No one should do what can be done by those who are younger or specialized in their field such as bee keepers, goldsmiths and crop cultivators. Each person has to work for his/her personal needs, for the needs of those who cannot work and for the society in general. Acquisition of wealth for prestige, control of other people or power is immoral.[98] "To create wealth largely on a competitive basis, as opposed to a cooperative one" is immoral.[99]

As a matter of principle people are "expected to be in solidarity with one another especially during the hour of need."[100] Broodryk uses simple traditional terms to demonstrate the ideal of Ubuntu; that is, "If you have two cows and the milk of the first cow is sufficient for your own consumption, *Ubuntu* expects you to donate the milk of the second cow to your underprivileged brothers and sisters. You do not sell it: you just give it."[101] Caring is an important pillar in the *Ubuntu* worldview. [102] "One can say that *Ubuntu* ethics is anti-egoistic, as it discourages people from seeking their own good without regard for, or to the detriment of, other persons in the community."[103]

Metz notes that in the traditional societies south of the Sahara there is always and underlying and unwritten ethical principle that it is immoral "to distribute wealth largely on the basis of individual rights, as opposed to need."[104] This principle is

---

[95] Bujo (1992, pp. 35–36).

[96] Nyerere (1968); Hord and Scott Lee (1995, pp. 65–72).

[97] L. Magesa, *African Religion: The Moral Traditions of Abundant Life* (New York: Orbis Books, 1997), 277–8.

[98] Nyerere (1968, pp. 2–5).

[99] Metz (2007, p. 325).

[100] Mnyaka and Motlhabi (2003, p. 223).

[101] Johann Broodryk, *Ubuntu. Life Lessons from Africa* (Pretoria: Ubuntu School of Philosophy, 2002) p. 8.

[102] Broodryk, *Ubuntu. Life Lessons from Africa*, 2002, p. 48.

[103] Mnyaka and Motlhabi (2003, p. 224).

[104] Metz (2007, p. 326).

based on Ubuntu general principle of common human equality and communitarian understanding of human mutual need for each other. This Ubuntu ideal of distributive ethical principle can be summarized in the following phrase: From each according to ability; to each according to need.

2.2.1.3 Ubuntu Justice is Communitarian

Ubuntu ethics revolves around all that favors life. Each individual and the community as a whole have a sacred duty to promote life. To underline the duty of the community in promotion of life and the individual duty to support community Bujo simply states that in traditional sub-Sahara Africa, "Individuals live only thanks to the community...life is the highest principle of ethical conduct."[105] Onah notes that promotion of life is "the determinant principle of African traditional morality and this promotion is guaranteed only in the community."[106] Metz notes that African respect for personal dignity is expressive of its respect for the sacredness of human life.

Metz articulates with clarity one of the cardinal principles of Ubuntu ethics. Basing his argument on Shutte's work Metz states that "An action is right just insofar as it positively relates to others and thereby realizes oneself; an act is wrong to the extent that it does not perfect one's valuable nature as a social being."[107] This statement explains the communitarian nature of Ubuntu justice. Justice is a socio-ethical principle which guides human interaction and relationships. The principle also entails the fact that self-realization happens within the communitarian setting. The starting point of a moral act is 'other-oriented.' Moral action should not infringe on the rights of others. In Metz's words, "an act is right if and only if it develops one's social nature without violating the rights of others."[108] This principle is necessary for community life.

Just action is that which facilitates or enhances personal realization. However, individual realization can only happen in the context of community. Moreover, self-realization should be both for self and for other related humans. Actually, Ubuntu contends that human self-actualization happens through other humans, which means that it cannot happen in isolation. Ubuntu justice is based on the identity of the self which is always inter-subjective, thus contingent to community. This phenomenon is best described by Seyla Benhabib. She states that "Individuation does not precede association; rather it is the kind of associations which we inhabit that define the kinds of individuals we become."[109] In other words society precedes an individual, defines the individual and helps the individual to self-realize.

The individual is a product of the community and owes his existence to the community. There is mutuality of responsibility, duty and rights between the community

---

[105] Bujo, *Foundations of an African Ethic: Beyond the Universal Claims of Western Morality*, 2001, p. 2, 3, 4, 52, 62, 66, 88.

[106] Godfrey (2012).

[107] Metz (2007, p. 331).

[108] Metz (2007, p. 332).

[109] Benhabib (1997, p. 73).

and its members. Such mutuality is based on individuals' neediness of the community for survival. Using Mbiti's words, the "community must therefore make, create or produce the individual; for the individual depends on the corporate group."[110] Mbiti explains, "Nature brings the child into the world, but society creates the child into a social being, a corporate person, for it is the community which must protect the child, feed it, bring it up, educate it and in many other ways incorporate it into the wider community."[111]

Consequently, the child has an obligation to live in such a way that his individual rights nurture and enhance the existence and flourishing of the community which enables not only the possibility of such rights but more importantly human individual life.

Community building is represented and expressed in almost all important activities of an individual or family. Everybody should play a role in nurturing community bonds. There cannot be a completely exclusive individual right. Among the Chagga and Setswana society, for example, slaughtering an animal and consuming it with the immediate nuclear family without giving rightful portions to members of the extended family, however little the piece meals may be is considered immoral. It is equivalent to theft.[112]

While from the western perspective there is naturally no entitlement in what one does not own, among the Bantu people, the entitlement is validated by the duty of each person to build the necessary bonds which foster and nurture community building. By being a member of the community everybody has a valid claim to what maintains the bonds without which the community cannot survive.

In sum, Ubuntu justice is essentially and always communitarian. Metz sums up Ubuntu ethics of communitarianism by the moral principle he identified from his research in Ubuntu that "an action is wrong insofar as it fails to honor relationships in which people share a way of life and care for one another's quality of life, and especially to the extent that it esteems division and ill-will."[113] This perspective on justice is different from the popular tendency which focuses on justice from the perspective of individual rights and claims. Individual rights are only real in the context and matrix of community or society.

2.2.2 Diversity

Ubuntu respects human diversity. Diversity is beneficial to societal fulfillment; plurality enhances both personal and societal self-realization.[114] The culture of Ubuntu realizes the importance of diversity for personal self-realization as human beings, for societal prosperity and for moral living. This understanding is summarized in the

---

[110] Mbiti (1990, p. 107).

[111] Mbiti (1990, p. 107).

[112] Metz and Gaie (2010, pp. 273–290).

[113] Metz (2009, p. 183).

[114] Broodryk (1997, pp. 5–7).

previously cited maxims that "a person is a person through other persons,"[115] and "a human being is a human being through the otherness of other human beings."[116] Van Der Merwe observes that Ubuntu dictates that to be human is to recognize the genuine otherness of fellow citizens. The recognition of and respect for each person's uniqueness is an essential component of society. This uniqueness involves the diversity of languages, histories, values and customs, all of which constitute human society.[117] This dissertation will explore in depth the need and respect for diversity in human society and ethical discourse in light of the culture of Ubuntu. As a result of the Ubuntu perspective of society as analogous to an organism, Ubuntu appreciates difference and diversity as richness. Diversity allows for variety of contribution to the community by each member for each member. Consequently, human society flourishes on diversity.

2.2.2.1 Anthropocentrism and Respect for Diversity

Most Sub-Saharan ethnic communities are radically anthropocentric. Bujo writes that "life is the highest principle of ethical conduct."[118] Everything revolves around the mystery of human life. Human life is so important that everybody has to take responsibility to nurture it prior to birth and post mortem in form of the 'living-dead'. All human life, regardless of differences in color, ethnicity, wealth, and nationality is sacred. God is revered through human moral life. Consequently, there is not so much direct reference to God. Respect for any human life is considered an act of worship and reverence to God.[119]

In praxis, as Bujo well expresses it, "the living members of this 'mystical society'[120] have an inalienable responsibility for protecting and prolonging the life of the community in all its aspects."[121] Such responsibility extends to all humans. One should only be allowed to kill in self-defense. However different or unconforming human life is, it should be treasured and respected. No wonder Bujo notes that "the morality of an act is determined by its life-giving potential."[122] This respect for human life implies tolerance, patience and respect for diversity. Bujo observes, however, that "since the common good must have precedence over the individual good, an individual who is really a danger for the community, or threatens the clan with loss of life or goods, must be simply removed."[123] However, "the main goal of African ethics is fundamentally life itself." The community is at the service of each life.[124]

---

[115] Shutte (1993, p. 46).

[116] Van Der Marwe and Willie (1996, pp. 1–3).

[117] Van Der Marwe and Willie (1996, pp. 2–3).

[118] Bujo (2001, p. 3).

[119] Bujo (1992, pp. 17–37).

[120] Bujo (1992, p. 22).

[121] Bujo (1992, p. 22).

[122] Bujo (1992, p. 22).

[123] Bujo (1992, p. 34).

[124] Bujo (2001, p. 2).

2.2.2.2   Otherness as Source, Objective and Rationale of Morality

Even though Ubuntu is basically Unitarian, diversity is an important part of it. Diversity belongs to the very essence of Ubuntu. It is the diversity that underlies the importance of unity. One of the maxims most expressive of the core meaning of Ubuntu and which has been discussed earlier in this work underlies importance of diversity for any meaningful community and individual social, moral, and psychological development. The differentness of others helps people recognize their own uniqueness, role, importance, duty and neediness."[125] The differentness of others includes diversity of languages, histories, values and customs, all of which constitute human society.[126]

Mbiti writes that "in traditional life, the individual does not and cannot exist alone except corporately. He owes his existence to other people, including those of past generations and his contemporaries. He is simply part of the whole. The community must therefore make, create or produce the individual; for the individual depends on the corporate group."[127] Implied in Mbiti's statement is the fact that the community helps the individual become different and unique while at the same time instilling in him or her communitarian accepted moral norms and ideals.

Personhood is a developmental concept in the culture of Ubuntu. Such development is facilitated by the community. Mbiti relates that, "Physical birth is not enough: the child must go through rites of incorporation so that it becomes fully integrated into the entire society." The initiation rites are usually age-related and vary depending on the specific ethnicity. According to Mbiti the rites signify moral, social, religious and behavioral development. "These rites continue throughout the physical life of the person, during which the individual passes from one stage of corporate existence to another. The final stage is reached when he dies and even then he is ritually incorporated into the wider family of both the dead and the living."

The dead members of the society remain living-dead until they are no longer remembered by any living person. They are believed to be constantly undergoing rites of incorporation into the world of the dead even as they are gradually forgotten by the living. Rites of initiation imply the role of the society in the work of creation. Mbiti elaborates this role when he writes that "Just as God made the first man, as God's man, so now man himself makes the individual who becomes the corporate or social man." Initiation rites need other people. Personal existence, completely independent of the society, is absurd. Thus Mbiti writes that "only in terms of other people does the individual become conscious of his own being, his own duties, his privileges and responsibilities towards himself and towards other people."[128]

In the process of individual formation by all other individuals and in all formal processes of initiation individual uniqueness is not only accepted or tolerated, it is cherished and given a special role in the society. The person is helped to know that he or she is unique, thus a needed organ within the community. Diversity is a bless-

---

[125] Van Der Marwe and Willie (1996, pp. 1–3).
[126] Van Der Marwe and Willie (1996, pp. 2–3).
[127] Mbiti (1990, p. 106).
[128] Mbiti (1990, p. 106).

ing to the community. To the individual, diversity and pluralism helps distinguish the self from the rest of the community members.

Initiation processes aim at cutting the umbilical cord continually so that the child is continually born into the wider human family, incorporated in it as his person-hood unfolds. One moves from one's mother into the nuclear family then extended family, then the ethnic group and then human family in general.[129] Mbiti writes that those initiation rites have great formational and educational purposes. "The occasion often marks the beginning of acquiring knowledge which is otherwise not accessible to those who have not been initiated. It is a period of awakening to many things, a period of dawn for the young. They learn to endure hardships, they learn to live with one another, they learn to obey,"[130] to mention but a few things.

Initiation, therefore prepares the candidates to deal with, accept, and use diversity for the common good. The continual rites of initiation aim at helping the youth, to accept their role in the wide human society, honor, respect and nurturing of every human life. One of the most important tests in the rites is a lesson of accepting diversity and using it for both communal and self-benefit.

Just as an individual cannot survive without the support of other individuals and the community at large, Ubuntu believes that no community can survive in the cosmos alone without being in solidarity with the rest of communities which share the earth. Diversity and uniqueness, both among individuals and among societies is riches, especially because, according to Ubuntu, humanity is, by large a product of human relationships. This world-view is seen in Ubuntu's emphasis on establishment and maintenance of harmony between different ethnicities.

2.2.2.3 Tension Between Diversity, Communitarianism and Human Freedom

According to Mbiti there can neither be freedom nor real ethical existence independent of the community. Mbiti states that individuality is based on plurality, in the sense that among the Bantu peoples of the sub-Saharan Africa individual existence is based on communal existence. This is a major contention in ethics of individual rights, since such ethics does not necessarily view individual existence as contingent to communal or societal existence; at least it does not emphasize the role of the community as Ubuntu does. Ubuntu communitarian ethics is based on the indebtedness of any particular individual both to the current community and to his ancestors who are responsible to who any particular individual becomes.[131]

Mbiti's interpretation of Ubuntu worldview reveals tension between individual autonomy, which is necessary for real human freedom, and Ubuntu communitarianism which is *sine qua non* of individual existence. Since the community defines the individual and that it takes precedence over individual personal autonomy and

---

[129] Mbiti (1990, pp. 118–129).
[130] Mbiti (1990, p. 119).
[131] Mbiti (1990, p. 106).

liberty, individual existence is only significant within the confines of the community. Obviously, Ubuntu's understanding of individual identity as interpreted by Mbiti resonates with Taylors' but it goes much further. According to Taylor, one's identity is not worked out in isolation. It is a work in progress, a negotiation through dialogue "partly overt, partly internalized, with others." Self-identity, therefore, cannot be independent of others or society.[132]

Mbiti posits that "the community must therefore make, create or produce the individual...Physical birth is not enough: the child must go through rites of incorporation so that it becomes fully integrated into the entire society."[133] This later statement reveals yet another difficult tension between Ubuntu respect for diversity and Ubuntu communitarianism. If the society produces the individual through continual initiations throughout life, one could validly question Ubuntu's tolerance of diversity and pluralism within and outside the community. However, Ubuntu does not only nurture diversity, it encourages diversity provided that it doesn't threaten communal existence. Communal existence is the measure of morality of a human act in Ubuntu.

The ideal of Ubuntu ethics is moral identification of an individual and the community. The approach Mbiti uses can be simplified by analogy of an organism. Since the community and the individual are one, whatever hurts the individual hurts the community and whatever hurts the community hurts the individual just as whatever hurts any part of an organism hurts the whole organism and whatever hurts the whole organism hurts all its parts. To be cut off from the community is tantamount to homicide since "to be is to belong." Interpreting Mbiti's perspective of Ubuntu, Chachine[134] states, since "to 'be' is to 'belong,' therefore to separate the individual from his social existence is to deny the individual the very freedom he seeks."[135] Interpreting Mbiti's perspective on freedom Chachine writes, "One cannot extricate the individual from his or her social environment without harming the very foundations of his or her freedom; without undermining the very social surroundings where he or she belongs."[136]

This statement means that moral life require human freedom, while human freedom is limited by the community or society in which a person is a member. "So to understand the context of the self is equivalent to understanding what one's freedom entails or should be."[137] Consequently, freedom is a relative term whose definition is provided by the community. The self being part of its social environment, "the ideal of freedom which may follow is that of 'situated' freedom as contrary to the

---

[132] Charles Taylor, *The Ethics of Authenticity* (Harvard University Press, Cambridge, Massachusetts 1991), p. 47.

[133] Mbiti (1990, p. 106).

[134] Chachine (2008, p. 233).

[135] Chachine (2008, p. 233).

[136] Chachine (2008, p. 233).

[137] Chachine (2008, p. 233).

idea of freedom as autonomy, 'choice', or self-determination. Therefore, the ideal of social solidarity is a central concept in Mbiti's justification of freedom."[138]

Individual existence along with all its rights, duties, and responsibilities is absurd and unintelligible outside of the community since "in African terms, one's freedom is correspondent to one's ability to harmonize oneself with one's own social surroundings."[139] Traditional African communities' regard the self as an extension of the community and the community as an extension of the self. There can only be freedom to relate, not to dissociate. Dissociation from the community is fatal. Gyekye contends that the community and the individual should be ascribed the same moral status because the community cannot exist without the individuals who gives it its corporate existence while, at the same time, no individual could survive without the conducive supportive environment provided by the community. Gyekye concludes that "the most satisfactory way to recognize the claims of both communality and individuality is to ascribe to them the status of an equal moral standing."[140]

The process of helping a person deal with diversity and plurality starts at birth. Mbiti notes how the "placenta and umbilical cord symbolize separation of the child from the mother, but this separation is not final since the two are still close to each other."[141] The society has to help the child get into the process of gradually and continually belonging "to the wider circle of society... [It] begins to get away from the individual mother, growing into the status of being 'I am because we are, and since we are therefore I am.'"[142] Some traditional societies have a way of expressing this important symbolism ritually by, for example, throwing the placenta into the river, whose symbolic meaning is: "the child is now public property, it belongs to the entire community and is no longer the property of one person, and any ties to one person or one household are symbolically destroyed and dissolved in the act of throwing the placenta and umbilical cord into the river."[143]

The child grows away from its nuclear family into the wider world to embrace global pluralism and diversity without losing touch with its original circles of relationship. The more a person can recognize other persons as his equals, and address their needs with empathic understanding regardless their uniqueness, the more ethically mature that person is. In this way Ubuntu communitarianism is as well, and at the same time, pluralistic and universalistic.

Ubuntu meaning of freedom is different from the popular western meaning of freedom. Justification of human freedom in Ubuntu is absurd if it is does not involve the community or society. Chachine and Mbiti easily show why this is the case: if "to 'be' is to 'belong', this implies that to be 'free' is to 'relate.'"[144] This

---

[138] Chachine (2008, p. 233).

[139] Chachine (2008, p. 233).

[140] Gyekye (1997, p. 41).

[141] Mbiti (1990, p. 110).

[142] Mbiti (1990, p. 110).

[143] Mbiti (1990, p. 110).

[144] Chachine (2008, p. 233).

understanding of freedom is almost foreign to the popular understanding of freedom as detachment and non-relationship, if need be; or freedom as "self-mastery, the elimination of obstacles to my will, whatever these obstacles may be—the resistance of nature, of my ungoverned passions, of irrational institutions, of the opposing wills or behavior of others."[145]

The traditional concept of freedom is different from the understanding of freedom as equality. Freedom as equality means that individual humans are considered of equal moral standing and the society as of secondary moral standing. This perspective holds individual's dignity as much more important than any societal or corporal moral entities.[146] However, realistic freedom is always relational. Interpreting Mbiti, Chachine distinguishes freedom from liberty: "'I am, because we are; and since we are, therefore I am' inspires us to see freedom as tolerance and inclusion, it invites us to distinguish mere freedom from liberty, whereby freedom stands as being, as a natural endowment; since all human beings are born free." Chachine implies that realistic freedom involves personal relationships and engagements, since human beings are by nature relational and their realization is enabled by personal relationships with other humans.

In other words one cannot be humanly free if one does not have human relationships with other persons. Freedom thus understood, "stands as what a person is in the original stage; while liberty by being a process in itself it stands as a practical action into becoming, emerging in the context of social interactions, as one's capacity or attempt to become free." In Chachine's observation, therefore, liberty is a means to an end, which end is freedom. He states that Liberty "results in the context of human striving for freedom, in the context of one's attempt to become free or to become fully human."

Consequently, liberty is a process not an end in itself. Chachine explains that "ethically, in the Ubuntu conceptual moral scheme liberty, thus defined, emerges as our human attempt to move from *is* moral universe into *ought* moral platform." Thus, liberty is a fluid transitional term which "implies action into becoming." Its end is more freedom because "in the context of *is* it expresses what one ought to be, while in the context of act it illuminates what one ought to do."[147]

Freedom however is an end, not a means. Human growth and development aims at greater freedom. However, freedom does not exclude human need for, and capacity to relate. According to Mbiti "what gives our lives meaning and purpose is our belonging and our capacity to exercise our own freedom in the realm of our human commitment and relationships."[148]

Ubuntu freedom is consistent with Temple's description of freedom. He states that freedom may be justified "only when it expresses itself through fellowship; and

---

[145] Isaiah Berlin, *Four Essays on Liberty* (Oxford University Press, Oxford, 2002), p. 193.

[146] MacIntyre (1984, p. 250).

[147] Chachine (2008, p. 234). In this passage Chachine cites and interprets Mbiti's distinction of freedom from Liberty.

[148] Chachine (2008, p. 234). In this passage Chachine cites and interprets Mbiti's distinction of freedom from Liberty.

free society must be so organized as to make this effectual; in other words it must be rich in sectional groupings or fellowships within the harmony of the whole."[149] Ubuntu integrates and weaves together communitarianism, diversity and freedom as the ideal of morality.

There is no question that Africa is a composition of many unique cultures and languages; however, one can rightly speak of a common African culture, the unifying culture that underlies all the unique different sub-cultures.[150] Tangwa refers to this synthesizing ability of Ubuntu and similar African cultures when he states that African cultures are "characterized by diversity and, left to themselves, united in their tolerance and liberalism, live and let live attitude, non-aggressivity, non-proselytizing character and in their accommodation of the most varied diversities and peaceful cohabitation of the most apparently contradictory elements."[151]

2.2.3 Biosphere

Ubuntu calls for respect of the biosphere. The cosmos has an inherent hierarchy of rights on which human rights are based. Every society and individual has an obligation to promote and protect the rights of the biosphere.[152] The culture of Ubuntu respects and reverences the integrity of the cosmos which supports the biosphere and human society. Dona Richards expresses this Ubuntu attitude toward the cosmos when she states that exploitation of the cosmos is self-defeating.[153] Richards notes that there is harmony in nature that should be respected as a matter of justice.[154] Since religion permeates all aspects of life in the culture of Ubuntu, there is no formal distinction between the sacred and the secular, between the religious and non-religious, between the spiritual and the material areas of life.

Likewise, morality permeates all aspects of life and environment. It matters how one treats wildlife or even non-living parts of creation. Violence towards anything inevitably meets a violent reaction.[155] It can be concluded that Ubuntu encourages a view of human life that is not independent of the biosphere, ecosystem and the cosmos. Ubuntu realizes that there is a network of interdependence without which individual and societal human life is impossible. Since the biosphere and the cosmos sustain human society, the society should preserve the integrity of the biosphere and the cosmos.

---

[149] William Temple, *Christianity and Social Order* (London: SCM Press Ltd, 1950), p. 65.

[150] Godfrey B. Tangwa, Elements *of African Bioethics in a Western Frame* (Mankon, Bameda: Langaa Research and Publishing Common Initiative Group, 2010), 12.

[151] Tangwa (2010, p. 11).

[152] Tempels (1946).

[153] Richards (1980, pp. 76–77).

[154] Richards (1980, pp. 76–77).

[155] Mbiti (1990, p. 1).

Consequently, Senghor notes that African culture conceives the world beyond the diversity of its forms, as a fundamentally mobile, yet unique, reality that seeks synthesis.[156] This work enlightens this aspect of the culture of Ubuntu as useful for discerning ethical concerns when applied to modern trends in global bioethics regarding pollution, climate change, extinction of some species, and the human role in the destruction of the biosphere.

2.2.3.1 The Self and the Cosmos in Relationship

In order to understand the indigenous African conception of reality, causality and the network of relationships between realities, one has to study the work of Placide Tempels[157] and his idea of 'force.' Even though some scholars have criticized Tempels' work and many have discredited it especially because of its exaggerated ambition, pride and generalization,[158] the work has a basic world view that is fairly representative and universal, at least to most indigenous African communities South of Sahara. In his view, Africans perceive and conceive of the world as a field of forces. Force is, in their view, nature of beings. Such forces are ordered hierarchically with God as the source of all force. God is the one "who has force, power, in himself. He gives existence, power of survival and of increase to other forces."[159]

Because all forces in their hierarchy of ability come from the same source, God, they are all related and interconnected. God enables all of them, consequently they are all related. In Tempel's words, "Created beings preserve a bond with one another, an intimate ontological relationship, comparable with the causal tie which binds creature and creator. For Bantu there is interaction of being with being, that is to say, of force with force."[160] According to Tempel the concept of force is metaphysical.

He observed that Africans perceive not only the empirical forces but their causality. He states that "Transcending the mechanical, chemical and psychological interactions, they [Africans] see a relationship of forces which we should call ontological...the Bantu sees a causal action emanating from the very nature of that created force and influencing other forces."[161] Simply stated being or existence is perceived as force and it all comes from and is sustained by God. It is all related although there is a hierarchy as per the kind of force and its influence on other forces.

The hierarchy of the forces is explained by J. Jahn who adapted the categories of A. Kagame.[162]

---

[156] Leopold Sedar Senghor, "Negritude: A Humanism of the Twentieth Century," in *I am Because We Are: Readings in Black Philosophy* (Amherst, MA: University of Massachusetts Press, 1995), p. 48.

[157] Tempels (1946).

[158] Mbiti (1990, p. 10).

[159] Tempels (1946, p. 61).

[160] Tempels (1959, p. 58).

[161] Tempels (1959, p. 58). The word in brackets is mine.

[162] Jahn and Kagame as cited in Mbiti (1990, pp. 10–11).

The categorization separates everything into basic four categories.

> *Muntu* is the philosophical category which includes God, spirits, the departed, human beings and certain tress. These constitute a 'force' endowed with intelligence.
> *Kintu* includes all the 'forces' which do not act on their own but only under the command of *muntu*, such as plants, animals, minerals and the like.
> *Huntu* is the category of time and space.
> *Kuntu* is what he calls 'modality', and covers items like beauty laughter etc.[163]

Mbiti proposes an ontology which is slightly different from Kagame's although it is equally anthropocentric. According to Mbiti there are five categories of being or forces:

> *God* as the ultimate explanation of the genesis and sustenance of both man and all things *Spirits* consists of extra-human beings and the spirits of men who died a long time ago *Man* including human beings who are alive and those about to be born. *Animals and plants*, or the remainder of biological life Phenomena and objects without biological life.[164]

The root—*ntu* is shared by all different kinds of forces and it represents force/being in general. Since being manifests itself only in particular beings. The root never appears without its manifestation as *Muntu*, *Kintu*, *Huntu* or *Kuntu* since it is the metaphysical being in itself or universal force. The universal force, however is the base of all force and by necessity relates all forces. No force can dissociate itself from it. Thus, reality is a unity which appears in a hierarchy of manifestations according based on the four categories mentioned above.[165]

Humans being are a force that is endowed with intelligence, freedom and autonomy. They are responsible for the necessary order and harmonious interaction of forces around them without detaching themselves from the lower forces in the hierarchy and the higher forces (elders, ancestors, spirits, divinities and ultimately God himself). Senghor explores how individuals in traditional African society are supposed to be responsible for ecosystems around them. Violence to nature was considered as violence to humanity, including the subject[166] since, as Sindima puts it, "nature and persons are one, woven by creation into one texture or fabric of life."[167] Consequently, the interests and wellbeing of an individual are subordinate to and dependent on the community and cosmic wellbeing.[168]

That is why Murove argues "that our human well-being is indispensable from our dependence and interdependence with all that exists, and particularly with the immediate environment on which all humanity depends."[169] To underline the direct relationship and symbiotic mutuality between an individual and the biosphere and the role of human individuals in within the cosmos Kasanane states that "An indi-

---

[163] Mbiti (1990, p. 11).

[164] Mbiti (1990, pp. 15–16).

[165] Mbiti (1990, p. 10).

[166] Leopold Sedar Senghor, "Negritude: A Humanism of the Twentieth Century, "p. 52.

[167] Sindima (1995, p. 127).

[168] Asante et al. (2008, p. 115).

[169] Murove (2004, p. 196).

vidual's good health is buttressed when he or she maintains environmental equilibrium, for instance, in the preservation of nature."[170] The interactive and symbiotic interrelationship between living beings and between the biosphere and the cosmos is fundamental in Ubuntu. The relationship is not only physical, biological and ethical; it is as well religious and eschatological. Writing about the role of a forest to human life, for example, Sindima states that "The forest provides the African with all basic needs—food, materials for building a home, medicine, and rain; it also provides a sanctuary for religious practices as well and a home for the fugitive; in addition, it serves as a cemetery and the abode of ancestral spirits." There is, therefore, recognition of the role and significance of nature in Ubuntu which calls for ethical responsibility on the part of humans who stand in constant need of the rest of biosphere and cosmos. With regards to the role of forest to Africans, Sindima concludes, "In short, the forest is everything for the African. It is this understanding of belonging to one texture of life which gives Africans the sense of respect and care for creation."[171]

Thus, while striving to promote and maintain both individual and societal wellbeing, indigenous Africans have always strived to attain and maintain personal and societal integration and equilibrium with their environment. They have always known that holistic human wellbeing is illusive if it excludes the environment which maintains it and without which human existence, live alone its wellness, remains an illusion. The environment is a partner and an extension of the individual and the community.[172]

2.2.3.2 Role of and Respect for Other Forms of Life

It must be stated that there is no treatise or consistent written account that explains the rationale behind most practices of African peoples south of Sahara. Most practice is based on unanimous understanding deducted from the nature of reality itself. Such understanding is based on the observation of cosmic interrelationships. Shutte observes that "Bantu psychology cannot conceive of man as an individual, as a force existing by itself and apart from its ontological relationships with other living beings and from its connection with animals or inanimate forces around it. The Bantu cannot be a lone being."[173]

He finds himself in a web of necessary ontological relationships with other beings including both past and future beings. His greatest value and objective is life itself. Consequently, Mbiti states that "average Africans see no need to enter into a

---

[170] Kasanene (1994, p. 350).

[171] Sindima (1995, p. 127).

[172] Kasanene (1994, p. 142).

[173] Shutte (1993, p. 55). Bantu people are the people who mainly share the Ubuntu worldview. They are basically indigenous south of Sahara. However, the world view is not limited to the Bantu. Peoples south of Sahara such as the Nilotes and Cushites share the worldview.

rational and theological squabble, to justify what they do, their concern is life and its wellbeing, how to protect and enhance it. 'Their philosophy of forces serves as sufficient guide'."[174]

Most indigenous peoples south of the Sahara believe that God created the world and established the order which humans discover. Human beings should respect the natural order as a matter of justice and respect for God. Nature serves human beings but injustice to it is punishable by God. For the Chagga, Akan, Ankore, Igbira, Kpelle and Illa, for example, the sun is central as a proof of God's providence in sustenance of living creatures for human beings. For the majority of African peoples rain is the most important expression of God's care for human beings. People like the Illa, Ngoni and Akamba hold that rain is the most important of the activities of God. When it rains God is generally happy with human beings. When there is drought, there is something amiss in people's relationship with God, especially in their treatment of nature.[175]

Ideally, the balance reflected in natural ecosystems should not be disturbed at all. Humans should limit the damage they inflict on animals and trees as much as they can. Food chains and the balance seen in habitats reflect God's wisdom and desire for order in creation. Human beings ought to respect it even as they have to fit into it and get their food from it. Destruction to nature should, therefore be minimal.[176]

Most African peoples South of Sahara believe in a real and organic relationship between humans and the land. Such relationship is usually expressed symbolically. Some Africans express this relationship by the burial of the placenta and the umbilical cord.[177] Some tribes plant the placenta with a seed of a fruit tree so that "as the person grows up, the tree also grows and he/she builds up a relationship with the tree. Since his/her umbilical cord has become part of the tree, the two (person and tree) are like brothers and sisters. Even if that person is to move far away there will always be a symbolic link of the invisible umbilical cord pulling the individual back to his/her homeland."[178]

The burial of the placenta and the umbilical cord serves as a covenant between the new-born child and the ancestral land. Exploring the relationship between land and African peoples, Ali Mazrui states that African attitude to land and nature in general is one of ecological concern and preservation. The "totemic frame of reference" is a caution against destruction or unjust exploitation of land.[179] Giles-Vernick observes that the solidarity between indigenous African peoples and nature is "mainly

---

[174] Mbiti (1990, p. 66).

[175] Mbiti (1990, pp. 40–49).

[176] Some (1998, pp. 49–50).

[177] Kamalu (1997, p. 161).

[178] Kamalu (1997, p. 161).

[179] Ali Mazrui, *Africa's International Relations: The Diplomacy of Dependence* (London: Heinemann, 1977), p. 265.

an acknowledgement of mutual interdependence."[180] The interdependence implies co-responsibility which on the part of humans includes restraint from "plunder of nature"[181] because it would hurt the human species.

According to Sindima by "interacting with nature, both creation and people give themselves a new meaning of life and through this relationship people discover themselves within the totality of all creation. As nature opens itself up to human-kind, it presents possibilities of experiencing life in its fullness. In the interaction with nature, people discover their being inseparably bonded to all life."[182] It consequently breeds a sense of *oughtness*, which is the source of ethical reflection. Thus, African people South of Sahara "conceive the world beyond the diversity of its forms, as a fundamentally mobile yet unique reality that seeks synthesis."[183] Ubuntu recognizes the unity of matter and its relationship with humans.[184]

Violence to land and nature is violence to the self and humanity in general. This is because of the intimate and necessary symbiotic relationship between humans and the biosphere in particular within inescapable cosmic context. Consequently, sub-Saharan Africans have a great sense or respect for the biosphere and the cosmos. Their view of human life is so holistic and inclusive that nothing is left out. There is interdependence, not only between human beings and their environment but also between material and spiritual aspects of reality.[185]

The relationship between human beings and their environment can be described as one of reverence. The reverence given to material reality is based on human need for it. Such reverence takes into account not just the current generations but, even more, future generations. Kamalu notes that respect and protection of material reality expresses a sense of responsibility for future generations and for the cosmos. It is about the survival of human species and other species in general. It "implies an ecological responsibility for the current generation of the living whereby the consequence of any actions for future generations must be considered."[186]

2.2.3.3   Sacredness of the Biosphere

Most indigenous people south of the Sahara view nature with deep reverence. It is "their first home, the home that holds the wisdom of the cosmos…Nature is profoundly intelligent as it stands, and human beings would do well to learn from its wisdom."[187] Some articulates how the sub-Saharan indigenous people respect order in nature. They believe that there is an on-going almost sacred wordless

---

[180] Giles-Vernick and Rupp (2006, pp. 61–62).

[181] Kinoti (1999, pp. 77).

[182] Sindima (1995, p. 126).

[183] Leopold Sedar Senghor, "Negritude: A Humanism of the Twentieth Century," p. 48.

[184] Nkrumah (1965, pp. 56–57).

[185] Sindima (1995, p. 127).

[186] Kamalu (1997, p. 158).

[187] Some (1998, p. 49).

communication between different creatures which should not be disturbed. Sustenance of ecosystems and food chains reveals part of nature's mind which should be kept sacred. Humans should never disrupt natural order. Nature sustains itself, regenerates itself and supports all it contains. Its integrity is sacred.[188]

Most Africans don't have to prove God's existence because; they have no problem perceiving God in their environment, leave alone believing that he exists. In their view nature manifests God. Mbiti observes that "all African peoples associate God with the sky or heaven…the majority thinks that He lives there; and some even identify him with the sky…among many societies the sun is considered to be a manifestation of God Himself and the same word or its cognate is used for both." This association of God with the sun is based on the centrality of the sun in the universe and its role in generation and sustenance of life. Mbiti cites some examples of such societies to be "the Chagga (*Ruwa* for both God and Sun), peoples of the Ashanti hinterland (*We* for both), Luo (*Chieng* for both), Nandi (*Asis* for God, *asita* for sun and Ankore (*Kazooba* for both)."[189]

Other African peoples such as the Elgeyuo, Ibbo, Suk and Tonga associate God with rain. Some trees, hills, rivers and caves are associated with God, thus regarded sacred.[190] Mbiti argues that for an indigenous African "nature is filled with religious significance…God is seen in and behind these objects and phenomena. They are his creation, they manifest Him, they symbolize His being and presence."[191]

Human psychic, emotional or physical disease results from either broken relationships with nature or with community. Human integrity and wellness cannot be conceived independent of nature and its principles and intelligence which is the context which is the base for all that is human. Thus Some argues that "our relationship to the natural world and its natural laws determines whether or not we are healed. Nature, therefore, is the foundation of healing…within the natural world are all of the materials and tenets needed for healing human beings."[192]

Some argues that human emotion is a door that connects humans with natural energy around them. Emotional energy communicates with natural waves of energy emitted by other beings in the biosphere. One should always learn to listen to the voice of one's emotions. Holistic healing should include emotional healing which ultimately grounds us with the biosphere.[193]

The indigenous peoples' ultimate meaning of illness is a breakage of relationship. "Some connection is loose or completely absent, or has been severed. What the villager sees in the physical illness is simply an aftermath of something that has happened on the level of energy or relationship."[194] This means that healing is a form of reconciliation, a "conjuring up an energy that will repair the spiritual state

---

[188] Some (1998, pp. 50–55).

[189] Mbiti (1990, p. 52).

[190] Mbiti (1990, pp. 52–56).

[191] Mbiti (1990, p. 56).

[192] Some (1998, p. 38.

[193] Some (1998, pp. 55–75).

[194] Some (1998, p. 73).

so that the spiritual healing can be translated into healing of physical disease."[195] In an attempt to bring about authentic healing one should know the proper herbs but, more importantly, one should know "the energetic background of the patient and the reason for the physical illness."[196] Moreover, the healer "has to go beyond the mere physiological and individual symptoms, until the proper psychological, moral and socially-conditioned cause can be traced and discovered."[197]

Human harmonious relationship with nature is of greatest importance since, as Some puts it, "when people die, nature is the only hospitable place where their spirits can dwell."[198] The dead maintain their relationship with the material world. They remember clearly the "experience of walking on the earth…the moments when they contributed to the greater good and helped to make the world better…they also remember with great remorse the failed adventures and the gestures that harmed others and made the world a less dignifying place."[199]

Indigenous people south of the Sahara have a holistic world view. The Dagara peoples, for example, have a cosmology which is inseparable from their psychology, ontology and eschatology. According to the Dagara, "matter and spirit are fused. The two phenomena are complementary, each a reflection of the other." The physical world we live in came into existence simultaneously with another world, a spiritual one which is more dynamic, expansive and much brighter. Each of the two aspects of reality, the material and the spiritual is manifestation of the other.

Humans are both spirit in form of matter. Some explicates the duality and mutuality of this cosmology in form of symbiosis. He states, "The connection to Spirit and the Other World is a dialogue that goes two ways. We call on the spirits because we need their help, but they need something from us as well…they look at us as an extension of themselves for their unrealized dreams which they can realize through us. They help us visualize and realize our own sacredness. We are looking up to each other and humans should take from this a sense of dignity."[200] In Mbiti's view, the "invisible world presses hard upon the visible and tangible world."[201] Although matter reflects the real reality, matter is a mere shadow of the reality; however, the real reality needs matter to express itself.

The Dagara view of reality is very similar to the platonic perspective of reality as a shadow of the ideal world; the world of ideas and concepts.[202] Mbiti views the spiritual and the physical as "two dimensions of one and the same universe. These dimensions dove-tail each other to the extent that at times and in places one is apparently more real than, but not exclusive of, the other."[203] Consequently, reality for

---

[195] Some (1998, p. 73).

[196] Some (1998, p. 74).

[197] Bujo (2001, p. 97).

[198] Some (1998, p. 54).

[199] Some (1998, p. 54).

[200] Some (1998, pp. 43–57).

[201] Mbiti (1990, p. 56).

[202] Some (1998, pp. 61–66).

[203] Mbiti (1990, pp. 56–57).

an African is essentially one; separation from the unity of nature which manifests spiritual unity of all that is in existence is annihilation. Ubuntu unity as proof of individual existence is thus demonstrated in the holistic worldview of the sub-Saharan indigenous peoples.

2.3   The Role of Solidarity

The third major component of the culture of Ubuntu emphasizes the role of solidarity. The meaning of this role will later be enlightened by considering the Roman Catholic ethical tradition. This component has three related concepts. First, pursuit of common good in every human action; Second, inculcation and maintenance of social cohesion; Third, minority empowerment for the sake of common good as a sign of ethical maturity.

2.3.1   Common Good

One of the most important objectives of Ubuntu is the pursuit of the common good for current and future human and non-human generations.[204] One of the qualities that differentiate Ubuntu from modern western ethics is that Ubuntu does not seek to promote the individual's interests more than it seeks to promote community interests and vice versa.[205] The culture of Ubuntu considers human action to be social. Every individual action has social implications and repercussions. Consequently, Symphorien Ntibagirirwa notes that Ubuntu arms one with "normative principles for responsible decision-making and action, for oneself and for the good of the whole community."[206] An ethically mature person is one who acts for common good. Such a person "can transcend, when necessary, what the customs of the family or the tribe require without disrupting the harmony and the cohesion of the community."[207]

The ethically mature person in the culture of Ubuntu does things not because they are required or expected but "because it is the right thing to do for both him/herself and the community."[208] This understanding will be paralleled with both Kohlberg's and Gilligan's theories of moral development. Ubuntu ethics considers any human act which ignores the common good to be unethical on the grounds that personhood is facilitated by, and dependent on, human society. Moral maturity implies awareness that one is a product of present and previous generations of human

---

[204] Nkrumah (1965, p. 59); Hord and Scott Lee (1995, p. 59).

[205] Nkrumah (1965, p. 59); Hord and Scott Lee (1995, p. 59).

[206] Ntibagirirwa (1999, p. 104).

[207] Ntibagirirwa (1999, pp. 104–105).

[208] Ntibagirirwa (1999, pp. 104–105).

community. Therefore giving back to the common good is a matter of justice rather than charity.

2.3.1.1   Common Ownership of the Major Means of Production

Indigenous African people fostered the common good. Common good is a contested phrase since it has been traditionally defined differently by different people. The nineteenth century individualist Jeremy Bentham defined it as "The sum of the interests of the several members who compose it."[209] Gyekye describes common good as "a good that is common to individual human beings—at least those embraced within a community, a good that can be said to be commonly, universally, shared by all human individuals, a good, the possession of which, is essential for the ordinary or basic functioning of the individual in a human society."[210] Gyekye further summarizes his description of common good as "that which inspires the creation of a moral, social, or political system for enhancing the well-being of people in a community generally."[211] It is Gyekye's understanding of common good that is employed in this work.

Indigenous sub-Saharan peoples resisted privatization of major means of production in order to safeguard the common good. Land, for example was almost always the property of all members of a given society. Everybody had a right of use according to the laws recognized by the society.[212] This was the community's way of ascertaining equality in acquisition and access to contribution to both the private and common good. Several post-independence African politicians interpreted this ethical regulation politically. They concluded that African traditional societies were socialist.[213] However, due to the fact that Ubuntu was an ethical culture which could not be reduced to political ideology, such politicians' ambitions failed.

To ascertain the decent minimum of survival requirements for all members of the society and to foster human dignity and security, Tangwa observes, "It was a taboo to sell or otherwise commercialize certain things, such as water, housing, fuel wood, the staple food, etc."[214]

In sub-Saharan Africa, human labor, as a means of production, has always been considered social and public. Although individuals retain their personal autonomy and private interests, there is a limit to the extent of private interest with regards to the outcome of their labor. The indigenous culture discourages extreme differences between the wealthiest and the poorest. There is a basic poverty line below which

---

[209] Jeremy Bentham, An *Introduction to the Principles of Morals and legislation*, Cited in Gyekye (1997, p. 45).

[210] Gyekye (1997, p. 45).

[211] Gyekye (1997, p. 46).

[212] Tangwa (2010, p. 147).

[213] Nyerere (1968, pp 1–12). Nyerere, Nkrumah and Senghor are some of the proponents of African socialism.

[214] Tangwa (2010, p. 46).

no one should be permitted to sink. There is also a ceiling line of wealth above which no one should go, relative to average individual and community wealth. Human labor is for private needs but within the limits and conditions set by the community so that it is for all as well.[215]

2.3.1.2 Distribution of Wealth on the Basis of Need

Sub-Saharan indigenous African societies are not socialist as many early post-independence African politicians argued.[216] Helping others is considered a moral requirement that cannot be overlooked. It is inconceivable to amass excessive wealth while fellow humans are in dire need. Amassing wealth for selfish reasons, regardless common good is considered a very dangerous sign in the unity and life of the society.[217] In the traditional society an individual who proved to be so selfish that he would accumulate wealth while others are in need of basic human needs would be considered as a criminal and an enemy of the community.

Distribution of wealth, was not forceful as is the case with political socialist approach, neither was it achieved through rhetorical persuasion. It rather happened naturally as an obvious moral requirement that everyone should observe. Wealth distribution aimed at attainment of the equilibrium that is considered by most sub-Saharan Africans to be an ethical ideal. In the words of Kasenene, "in all they do, Africans strive to promote the wellbeing of the members of society, and this is attained when there is personal integration, environmental equilibrium, social harmony, and harmony between the individual and both the environment and the community."[218] It is that equilibrium that will support life.

Because human life is of the greatest value in African morality, and the health of the biosphere is necessary for flourishing of human life, Mbiti notes that "indigenous Africans see no need to enter into a rational and theological squabble to justify what they do. Their concern is life, its wellbeing, how to protect and enhance it. 'Their philosophy of forces serves as sufficient guide'."[219] Distribution of wealth helps protect and enhance human life. The one who refuses to support life is an enemy of life, thus poison to the community and its survival.

The culture of Ubuntu had in place mechanisms to ascertain that every member of the society is enabled to employ his or her potential for the personal good and for common good. "In practice, if one had two cows for milk, he would donate one to a person who has none so that the person who has no cow would feed the cow loaned to him so that he can get a supply of milk for his family needs. Usually if the cow gets a calf, the first calf would belong to the owner of the cow and the second one would belong to the person feeding the cow, then the alternate cycle

---

[215] Metz (2007, pp. 325–326)

[216] Nyerere (1968, pp. 1–12).

[217] Metz (2007, pp. 325–326).

[218] Kasenene (1994, p. 142).

[219] Mbiti (1990, p. 66).

repeats itself. In that way, laziness is discouraged and every member of the society is enabled to participate both in personal and common good. For immediate need food and other basic needs such as food, water and shelter should be provided without hurting the human dignity of the recipient. No one can claim to be free from the plight of any other person in the community."[220]

The donation in this case is not charity but a duty. Refusal to donate is an ethical violation, especially if the poor party's life is jeopardized in any way. This example shows that Ubuntu is not a socialist ideology but a cultural ethic which values life. Ubuntu sharing aims at supporting all life by the community and each of its members. This perspective of Ubuntu is a great contribution to global bioethics and an element of constructive dialogue.

Distribution of wealth in sub-Saharan Africa is a practical application of the indigenous meaning and objective of justice as reparation and restoration. In many ways it is similar to Jewish understanding of justice as *tzedakah*. The word *tzedakah* literally means righteousness, charity, justice and obligation to the needy. In absolute terms the word is applicable to God only. "For the Lord your God, he is God of gods and Lord of Lords… he doth execute justice for the fatherless and widow and loveth the stranger, in giving him food and raiment" (Deuteronomy 10:19; 15:7–10; Psalm 132:15; 145:15–16). However, since human beings are created in God's image, they are challenged to be like God in holiness and justice. Actually, charity is analogous to lending to God as is indicated in Proverbs 19:17. In Judaism, as in sub-Saharan Africa, nothing really belongs to anyone. What is given to the poor, therefore, belongs to God and no human being has an absolute right to it.[221]

The Jewish scriptures reveal that justice is fundamental and a prerequisite if one is a believer or a member of society. If members of the society are just there will be no exploitation and each member will "enjoy at least a basic level of material security."[222] The poor, therefore, have a right and the rich have an obligation to give in *tzedakah* (charity) as a way of practicing justice. According to Jewish spirituality, "the poor man does more for the house holder (in accepting alms) than the house holder does for the poor man (by giving with charity)."[223]

The major difference between *tzedakah* and Ubuntu is that Ubuntu is neither enforceable nor does it have mathematical calculation of the exact amount to be given by each member of the community to the poor like *tzedakah* does. The second difference is that *tzedakah* does not limit one's possessions in relation to the average wealth of individuals in the community, that is, *tzedakah* does not have poverty line below which nobody is allowed to drop. The third important difference between Ubuntu and *tzedakah* is that *tzedakah* does not concern itself much about production. Ubuntu ethics compel every member of society to employ his potential and participate to the best of his ability and talent in the production of wealth for self and the community.

---

[220] Sisulu as cited in Metz (2007, p. 326).

[221] Mackler (1991).

[222] Mackler (1991, p. 225).

[223] *Tzedakah*, a Hebrew term, is generally drawn from the Jewish Scriptures. This section is from Lev. 34:8.

2.3.1.3 Moral Obligation to Participate in the Process of Production

Ascertaining common good is not based only on distribution it is important that everybody who can work does work. Nyerere notes that "in traditional African Society *everybody* was a worker. There was no other way of earning a living for the community. Even the elder who appeared to be enjoying himself without doing any work and for whom everybody else appeared to be working, had, in fact, worked hard all his younger days."[224] Thus the system was so organized that there is assurance that the elderly would be naturally protected as a matter of justice. Nyerere states that "the wealth he [the elder] appeared to possess was not his, *personally*; it was only 'his' as the elder of the group which had produced it. He was its guardian. The wealth itself gave him neither power nor prestige."[225]

Nyerere argues that traditional society had no room for an 'idler' or a 'loiterer.' It was an offence to the society not to work. The society was very hospitable to strangers and guests. However hospitality did not allow exploitation. To explicate this point Nyerere uses a common Swahili saying: "*Mgeni siku mbili; siku ya tatu mpe jembe.*—or in English, treat your guest as a guest for two days; on the third day give him a hoe!"[226] Usually, the guest would ask for the hoe long before his host is obliged by the demands of Ubuntu to hand him one.[227] Observing the traditional sub-Saharan African community, one finds embedded within it the principle of subsidiarity which enabled each member to be a participant according to his ability.

Nyerere notes that the traditional community strives to make sure that each person has the means to realize his potential both for the self and for the society.[228] Membership right (which is essential for survival as a person) in any given indigenous sub-Saharan community, cannot be separated from individual rights and responsibility for the good of the self and the community.[229] Consequently, there is mutual need between an individual and the community. Neither the community nor the individual can survive without the other.

It can be safely concluded that sub-Saharan indigenous African societies were "moderate communitarian" since, as Gyekye states, "the communitarian ethic acknowledges the importance of individual rights but it does not do so to the detriment of responsibilities that individual members have or ought to have toward the community or other members of the community…responsibility is an important part of morality."[230] Gyekye suggests ascribing the community and the individual in such a community "the status of an equal moral standing."[231]

---

[224] Nyerere (1968, p. 4).

[225] Nyerere (1968, pp. 4–5).

[226] Nyerere (1968, p. 5).

[227] Nyerere (1968, p. 5).

[228] Nyerere (1968, pp. 1–75).

[229] See Gyekye (1997, pp. 61–70).

[230] Gyekye (1997, p. 66).

[231] Gyekye (1997, p. 41).

2.3.2 Social Cohesion

Ubuntu fosters social cohesion.[232] Individual humans and the society as a whole exist in a symbiotic relationship. Each exists only in relationship with the other. The pursuit of the common good depends on all members of society recognizing of this relationship.[233] Since one becomes aware of one's own existence, duties, obligations and rights in and through, the community, Mbiti observes an implied but obvious bond between individuals so that when one suffers one does not suffer alone but one suffers with the whole group. The culture of Ubuntu views human society as an organism whose parts are all important for their contribution to the entire organism. That is why Mbiti argued that whatever happens to one affects the entire group; whatever happens to the group affects each member.[234] This reciprocal relationship between an individual and the community increases the sense of belonging. Mnyaka and Motlhabi affirm that in Ubuntu culture "Everyone belongs and there is no one who does not belong."[235]

Ubuntu is committed to upholding the values of the community. Community values are shared between "the living and their ancestors in a way that shows the living's commitment to fellowship with their ancestors and those values that have enabled them to live life in harmony with everything else in the community."[236] Social cohesion for the sake of protection, nurturing and fostering all human life is the ideal of Ubuntu.

2.3.2.1 Moral Responsibility to Participate in Community Building

Indigenous African people south of Sahara hold that it is a moral responsibility for members of communities to actively participate in all that contributes to the life of the community. Self-realization is undeniably dependent on the community. Individual self-realization is concomitant to, and in mutuality with community health. Consequently, Gyekye argues that, "the communal definition of constitution of the individual can only be understood in partial terms, requiring that both the individual and community be given equal moral worth."[237] Since individual life depends on communal life, one has to participate in the activities such as communal norms, rituals and traditions that contribute to the life of the community. Failure to do so is tantamount, not only to suicide but also to killing of the society. It is a crime.

---

[232] Tutu, *No Future Without Forgiveness,* In Michael Battle, *Ubuntu: I in You and You in Me*, 2009, p. 2.

[233] Nyerere (1968, pp. 1–12).

[234] Mbiti (1990, p. 1).

[235] Mnyaka and Motlhabi (2003, p. 222).

[236] Murove (2004, p. 200).

[237] Gyekye (1997, p. 52).

Personal behavior and conduct that upset integrity of the community is consequently immoral, therefore, to be discouraged.[238]

Normal interaction, spending time with others, and communication with one other in a community is not optional but a requirement for the life of the community. This societal obligation is best explained by a study done by Augustine Shutte. The study involves two groups of nuns in one convent: Africans and Germans. While the German nuns would continue engaging themselves in some materially productive activity after their daily chores, such as weaving and knitting, the African nuns spent a lot of time in conversations with one another. According to the study, each group blamed the other as morally lacking and irresponsible.[239] While the German nuns blamed the African nuns for wasting time and for being irresponsible, the Africans blamed the German nuns for caring more for their hobbies and practical matters than for people.

According to the African nuns it is unethical to not to engage others in maintaining and actively contributing life to the community. The German nuns did not see any sense in the mere lengthy *unproductive* talk among African nuns. This clash of cultures caused conflict based on different ethics. The German nuns failed to understand the significance of the dialogue between the African nuns. Its significance is in the very fact that it is not business oriented or geared towards any material gain. It was simply for the sake of community life in the sense of Ubuntu. This is best summarized by Ruch. For Africans living according to the ethic of Ubuntu, states Ruch, "What I am myself for and by myself, matters less than what I am with, in and through the others."[240]

The African nuns were there with, for, in and through their colleagues, and that is what really matters. Ideally their fulfillment is based on, not exclusive of, their confreres fulfillment. Life is all about participation and contribution in the rhythm of the community. Ruch says it in a very simple categorical statement: "to be is to participate."[241] According to Ubuntu participation is a moral ideal; failure to participate is an ethical omission.

While human dignity may be considered from an individualistic perspective, in Ubuntu human dignity is meaningless independent of the community. The role of community in recognition and ascertaining human rights cannot be exaggerated. Gyekye states that moderate communitarianism should not be obsessed with individual rights. "The communitarian society, perhaps like any other type of human society, deeply cherishes the social values of peace, harmony, stability, solidarity, and mutual reciprocities and sympathies." In Gekye's view such values are essential for existence of any real human community.

---

[238] Metz (2007, p. 327).

[239] Augustine Shutte, *Ubuntu: An Ethic for the New South Africa* (Cape Town: Cluster Publications, 2001), pp. 27–28.

[240] E. A. Ruch, "Towards a Theory of African Knowledge," in *Philosophy in the African Context*, ed. D. S. Georgiades and I. G. Delvare (Johannesburg: University of Witwatersrand Press, 1975), p. 18.

[241] Ruch, "Towards a Theory of African Knowledge," p. 10

He asserts that "in the absence of these and other related values, human society cannot satisfactorily function but will disintegrate and come to grief." In order that such values may be there, however, there is need for definition of individual limits. In Gyekye's words, "the preservation of the society's integrity and values enjoins the individual to exercise her rights within limits, transgressing which will end in assaulting the rights of other individual or the basic values of the community."[242]

It is the community which by recognizing one as human gives him his due respect as an equal and a participant in the life of the community. Mnyaka and Motlhabi state that a "person has dignity, which is inherent; but part of being a person is to have feelings and moral values that contribute to the well-being of others...it shows that one contributes to the definition of oneself through everything one does. One's identity or social status goes hand in hand with one's responsibility or sense of duty towards, or in relation to, others."[243] This means that human dignity is to be always understood in the matrix of the community.[244]

Kasanene explains this *status quo* at best when he writes, "one cannot regard even one's own life as purely personal property or concern. It is the group which is the owner of life, a person being just a link in the chain uniting the present and future generations."[245] The main contribution that this worldview illuminates to the global understanding of human dignity and human rights in general is the contingence of rights to community. It also highlights the responsibility of the individual to the community which prescribes the dignity due to any individual as human.

2.3.2.2   Respect for Personal Autonomy as a Requirement in Community Building

Teffo notes that due to the importance of social cohesion Ubuntu "discourages the view that the individual should take precedence over the community."[246] Ubuntu, however, does not suppress the individual's unique rights and privileges within the context of the community. Using the words of Macquarrie, Ubuntu "preserves the other in his otherness, in his uniqueness, without letting him slip into the distance."[247] In other words Ubuntu defines, respects, and promotes personal autonomy within the limits of common good. Common good is severely damaged if self-determination is not honored by the community. However, membership in the community is *sine qua non*. Bujo explains this when he states "—individuals live only thanks to the community."[248] Mbiti provides an explanation of the statement of Bujo when he writes "in traditional life the individual does and cannot exist alone

---

[242] Gyekye (1997, p. 65).

[243] Mnyaka and Motlhabi (2003, p. 224).

[244] Gyekye (1997, p. 63).

[245] Kasanene (1994, p. 349).

[246] Teffo (1994, p. 7, 12).

[247] Macquarrie (1972, p. 110); Shutte (1993, p 49, 51).

[248] Bujo (2001, p. 1).

except corporately. He owes his existence to other people, including those of past generations and his contemporaries. He is simply part of the whole." In other words self-hood does not develop entirely from within a person. Its stimulus is outside the person.

It also means that a person is really a product of both his or her current human society and preceding generations. "The community must therefore make, create, or produce the individual; for the individual depends on the corporate group." Although Mbiti does not mention it for the sake of emphasizing the role of the community in personal formation, reciprocity is essential in the process. Mbiti emphasizes that "physical birth is not enough: the child must go through a rite of incorporation so that it becomes fully integrated into the entire society."[249] However, the child in the initiation process retains his or her uniqueness and autonomy as a person.

The child responds and reciprocates to the community by becoming a unique, proactive and productive member for the sake of the self and for common good. Michael Battle elaborates the same argument provided by Mbiti when he writes "We say a person is a person through other persons. We don't come fully formed into the world…we need other human beings in order to be human. We are made for togetherness; we are made for family, for fellowship, to exist in a tender network of interdependence."[250]

Being preceded by the community and being dependent on it for his survival, the individual needs the community just as the community needs the individual. To explain this fact Kwame cites Akan saying "When a human being descends from heaven, he [or she] descends into a human society." So the person should not live in isolation from other people since part of his constitution comes from inevitable social relationships, without which self-realization is impossible. There is, therefore, a delicate balance between individual self-determination and the context in which it is practiced, which context is the community.

Regarding this delicate balance Gyekye states "It might be thought that in doing so, such an arrangement tends to whittle away the moral autonomy of the person—making the being and life of the individual totally dependent on the activities, values, projects, practices, and ends of the community… that arrangement diminishes his freedom and capability to choose or re-evaluate the sheared values of the community."[251] However, as John Macquarrie, writes in *Existentialism,* when communitarianism becomes oppressive, then Ubuntu is abused. Ubuntu respects individual autonomy, "true Ubuntu incorporates dialogue.

It incorporates both relation and distance." Ubuntu maintains personal autonomy without encouraging individualism.[252] Ndaba makes this important point clear when he argues "that the collective consciousness evident in the African culture does not mean that the African subject wallows in a formless, shapeless or rudimentary collectivity." On the contrary it "simply means that the African subjectivity develops

---

[249] Mbiti (1990, p. 106).

[250] Battle (1997, p. 65).

[251] Gyekye (1997, pp. 36–40).

[252] Macquarrie (1972, p. 110); Shutte (1993, p. 49, 51).

and thrives in a relational setting provided by ongoing contact and interaction with others."[253]

Although there is an inclination towards collectivism and a sense of communal responsibility in the philosophy of Ubuntu, individuality is not negated but affirmed in interpersonal relationships within the society. The 1997 South African Governmental White Paper for Social Welfare officially recognized Ubuntu as "the principle of caring for each other's well-being" It called it a "principle of mutual support."[254] Mutual support is not contradictory, but supportive of individual identity and autonomy. Teffo explains that Ubuntu "merely discourages the view that the individual should take precedence over the community."[255] Furthermore mutual neediness within community members is crucial as Broodryk explains. He posits that as a process of self-realization *through* others, Ubuntu enhances the self-realization *of* others.[256]

Realistically, Ubuntu recognizes the importance of human relationship without which autonomy cannot be comprehended. John Macquarrie explains that in Ubuntu individuals can only exist as human beings in their relationship with other humans. The word "individual" therefore, "signifies a plurality of personalities corresponding to the multiplicity of relationships in which the individual in question stands." Hence, "being an individual by definition means 'being-with-others.'"[257]

Weil affirms that Ubuntu champions realistic freedom; that is, "it is not true that freedom of one man is limited by that of other men." Freedom is always relative to the freedom of others. "Man is really free to the extent that his freedom fully acknowledged and mirrored by the free consent of his fellow men finds confirmation and expansion of liberty. Man is free only among equally free men." Ubuntu recognizes the fact that "the slavery of even one human being violates humanity and negates the freedom of all."[258]

Due to indigenous Africans' rootedness into community as the only way to survive and grow as an individual, colonialism and neo-colonialism had not only a political impact on indigenous African communities, but had also psychological, social, ontological and ethical impact. Mbiti refers to African situation after colonialism when he writes "modern change has brought many individuals in Africa into situations entirely unknown in traditional life…The change means that the individuals are severed, cut off, pulled out and separated from corporate morality, customs and traditional solidarity. They have no firm roots anymore."

One of the worst legacies of colonialism consists of taking a people from the culture and ethics that define them without replacing it with another. Mbiti describes such situation in a dramatic way. He says, "They are simply uprooted but not

---

[253] Ndaba (1994, p. 14).

[254] South African Governmental White Paper for Social Welfare, http://www.welfare.gov.za/Documents/1977/wp.htm. February 15, 2012.

[255] Teffo (1994, p. 7, 12).

[256] Broodryk (1997, pp. 5–7).

[257] Macquarrie (1972, p. 104).

[258] Weil (1973, p. 182, 188–189).

necessarily transplanted. The traditional solidarity in which the individual says 'I am because we are, and since we are, therefore I am', is constantly being smashed, undermined and in some respects destroyed." Colonialism imposed not only political and economic control over the peoples of Africa; it imposed a foreign culture which was opposite the traditional culture and ethics of Ubuntu. Mbiti noted that at his time emphasis was "shifting from the 'we' of traditional corporate life to the 'I' of modern individualism."[259]

In sum, personal autonomy is essential in Ubuntu caring since in its absence neither caring nor community is possible. Ubuntu forms persons to be autonomous, although always within the limits of what is acceptable by the society, since there cannot be real individual human existence outside human community. Personal autonomy in Ubuntu, therefore, is logically and simultaneously for the good of the self and for common good. In addition to its illumination on the necessity of human relationships, which are facilitated by the implied personal autonomy, Ubuntu reinforces the role of human society, which formulates principles of ethics, as indispensable.

2.3.2.3 Community as an Extension of the Individual

Ubuntu social cohesion is an expression of care that is essential for the existence of the human community as a whole and for each individual in it. It is the kind of care advocated by most care ethicists. Ubuntu social cohesion means assumption of responsibility and active participation in the community for self-realization and for other people's realization. For this reason Ubuntu culture fosters a feeling of integration between individuals and their society. The society is almost regarded as an extension of the self in the sense that whatever is done by any member of the society affects each other member of the society and the society as a whole. Such understanding fosters regard for responsibility, duty and care.

Due to its communitarian mindset, indigenous sub-Sahara African communities represented by Ubuntu world view define individuality by a different criterion from the popular western criterion. "It is not an individual *vis-à-vis* (against) community but an individual *a la* (with) community. It is pro-community rather than against community."[260] This mindset promotes a caring attitude. Caring for one's neighbor and community means taking part in all communal and neighborhood activities, and caring is crucial in the culture of Ubuntu. One is naturally "expected to be in solidarity with one another especially during the hour of need."

That kind of solidarity is clearly manifest in events such as death. Neighbors would spend hours, sometimes days with the bereaved family as a way of alleviating their pain and strengthening them.[261] Munyaradzi observes that in traditional African ethics, a patient would not go the doctor alone. He would usually be

---

[259] Mbiti (1990, pp. 219–225).

[260] Macquarrie (1972, p. 104).

[261] Mnyaka and Motlhabi (2003, p. 223).

accompanied with his or her relatives and neighbors. The company of relatives and neighbors helps to provide for the needed support, counseling, interpretation and understanding of both the diagnosis and prognosis.[262] Munyaradzi's observation is one of many illustrations which helps explore the communitarian and Unitarian ethics of Ubuntu.

Simply put, the analysis means that there is no absolute secrecy. The communitarian nature of the culture of Ubuntu cannot allow the separation caused by the demand for privacy that modern medicine would expect. In fact, in some instances, the doctor would avoid giving the detail of the diagnosis of a patient directly to the patient while revealing it to family. Often times this happens to protect the patient from the pain of dealing with the bad news while, at the same time helping the family help the patient coup.

Ubuntu can rightly be said to be at least minimally moderate communitarian. Gyekye describes moderate communitarianism as "a model that acknowledges the intrinsic worth and dignity of the individual human person and recognizes individuality, individual responsibility and effort."[263] Ubuntu, however, is much more communitarian than moderately so. Senghor describes African communitarianism more elaborately when he states that among Africans community and community activity takes precedence over individuals and their individual activity without disregarding or underrating the importance of each individual, for himself or herself and for the community.[264]

Ubuntu therefore is essentially and inescapably communitarian. Gyekye explains, "Communitarianism immediately sees the human person as an inherently communal being, embedded in a context of social relationships and interdependence, and never as an isolated, atomic individual."[265] The Bantu people help to explicate this in their casual conversational language. Nussbaum notes how the Shona people of Zimbabwe, for example, have this morning greeting: "*Mangani, marara sei?* (Good morning, did you sleep well?" The response is always: "*Ndarara, kana mararawo.* (I slept well, if you slept well)."[266] Mbombo writes about how an individual from the country would "go to town, to tell us the whole story of their illness and how somebody else is not well in the family, and how somebody is not well in the community."[267] Broodryk notes the same mindset in the greeting "*ninjane*" which represents not just an inquiry about personal well-being but also about the well-being of the subject's relatives, friends and neighbors.[268] Sanon observes that

---

[262] Murove (2005, p. 146).

[263] Gyekye (1997, p. 40).

[264] Senghor (1964, pp. 93–94).

[265] Kwame Gyekye, "Person and Community in African Thought," in *Philosophy from Africa. A Text with Readings* ed. P. H. Coetzee and A. P. Roux (Durban: International Thomson Publishing Southern Africa, 1998), p. 319.

[266] Nussbaum (2003).

[267] O. Mbombo, "Practicing medicine across cultures: conceptions of health, communication and consulting practice." In M. Steyn and K. Motshabi, eds. *Cultural Synergy in South Africa. Weaving Strands of Africa and Europe* (Randburg: Knowledge Resources, 1996), p. 110.

[268] Broodryk, *Ubuntu. Life Lessons from Africa*, p. 101.

"Where a European may only inquire after the health of someone he meets, the African wishes to know, even from a total stranger, whether his family members are well.

Not only a 'How are you?' is important, but rather, 'How are your people?' is decisive in regarding health."[269] There is no doubt, therefore, that communitarianism is at the heart of indigenous African way of life, so much so that immediate community is viewed an extension of the self. This state of affairs is based on what Mbiti observed, that is, "the individual in African tradition does not and cannot exist alone, but that he or she exists corporately, such that they owe their existence to other people."[270]

Thus Ubuntu is about intrinsic connectedness of humanity. Using an analogy of a swimmer and the sea Ruch explores African perspective on life as that of interconnectedness.[271] In Ubuntu culture life is participation of an individual in the life of his or her community, in the eco-system, and in the cosmos even as the human community, the biosphere and the cosmos participates in the life of each individual. Thus life is about connectedness and participation. Individuals recognize the life of the community and affirm it in its riches; the community recognizes the life of each individual in it and affirms it in its uniqueness.

No one is exempt from Ubuntu communitarianism since there is no life outside it. Consequently Macquarrie observes that "being with others…is not added on to a pre-existent and self-sufficient being; rather, both this being (the self) and the others find themselves in a whole wherein they are already related. By nature a person is interdependent with other people."[272] Realization of human interdependence commands what Teffo calls "respecting the *historicality* of the other. Respecting the historicality of the other means respecting his/her dynamic nature or process nature."[273] Consequently, notes Tutu, a person who embraces Ubuntu is "open and available to others, affirming of others, does not feel threatened that others are able and good, for he or she has a proper self-assurance that comes from knowing that he or she belongs in a greater whole and is diminished when others are humiliated or diminished, when others are tortured or oppressed."[274] In brief, the community is an extension of the individual; ideally, the individual must see himself or herself in the community in whose existence he shares.

---

[269] A. T. Sanon, "Heil und Heilung fur den Christen in Afrika," cited in Bénézet Bujo, *The Ethical Dimension of Community: The African Model and the Dialogue between North and South* (Nairobi: Paulines Publications Africa, 1998), pp. 182–183.

[270] Mbiti (1997, p. 141).

[271] E. A. Ruch, "Towards a Theory of African Knowledge," in *Philosophy in the African Context*, ed. D. S. Georgiades and I. G. Delvare (Johannesburg: University of Witwatersrand Press, 1975), p. 2.

[272] Macquarrie (1972, p. 104).

[273] Teffo (1994, p. 11).

[274] Tutu (1999) cited by Battle (2009, p. 2).

2.3.3   Minority Empowerment

Ubuntu supports minority empowerment. Minority recognition, protection, enablement and empowerment for the sake of the common good are measures of a specific
community's ethical maturity.[275] Minority empowerment in Ubuntu is not just a
matter of charity, or a religious practice, it is an ethical imperative which defines a
person and society at large. Mutual, peaceful co-existence with decent minimum for
all is an inevitable ideal of life since there is no separation between human rights,
religion, ethics and other aspects of life.[276] Ubuntu culture opposes the individualism that Naomi Scheman considers repulsive due to its marginalizing effect on the
minority.[277]

2.3.3.1   Minority Empowerment as Defense of Basic Human Right
          to Life and Dignity

Sub-Saharan indigenous African communities have the concept of, and have been
living according to human rights based on human dignity. Sundman defines right as
"a legitimate claim and corresponding duties."[278] Sundman further defines human
right as a "right which human individuals have simply by virtue of being human."[279]
Generally, "human rights protect the value of welfare, but only to the extent that this
corresponds with our authentic needs."[280] The fact that human rights incorporate
both legitimate claim and corresponding duties, implies that human rights are based
on human reciprocity. Thus human rights result from human relationships within
society.

Since indigenous sub-Saharan Africans "do not think in 'either/or,' but rather in
'both/and' categories,"[281] their concept of human rights appears to weigh more on
the side of duties of the society and its members rather than on the claim of an individual. The claim is implied in the duties because, as Ruch puts it "myself, matters
less than what I am with, in and through the others…Existence is not merely 'being
there;' it is power of participation in the pulsation of life. 'To be is to participate.'"[282]

The sub-Saharan concept of human rights revolves around human life. Bujo observes that "the community must guarantee the promotion and protection of life
by specifying or ordaining ethics and morality."[283] The indigenous preoccupation

---

[275] Nyerere (1968, pp. 1–12).

[276] Kasanane (1994, p. 140).

[277] Scheman (1983, p. 240).

[278] Sundman (1996, p. 183).

[279] Sundman (1996, p. 183).

[280] Sundman (1996, p. 183).

[281] Bujo (2001, p. 1).

[282] Ruch, "Towards a Theory of African Knowledge," p. 10.

[283] Bujo (2001, p. 2).

with human life has led some scholars to misjudge Africa to be too anthropocentric and communitarian to have a clear separation of claims from duties or ethics from religion.[284] Bujo states that in the past some scholars have argued that a person in Africa "is ethically subsumed under ethnic group to such an extent that he scarcely merits to be considered as an autonomous ethical subject."[285] If this were the case, it would be impossible to speak of individual human rights.

Bujo observes that recent research, however "has proven conclusively that the group does not at all dissolve the ethical identity of the individual. This is confirmed in a number of proverbs."[286] Consequently, Africans do have human rights. Actually, the community is at the service of each human life with its uniqueness as an irreplaceable organ of the community. At the same time, the role of community in ethical conduct and human individual human rights is indispensable. Cut off from human community, the individual loses personhood along with all its rights and privileges.[287]

In the culture of Ubuntu, the basis and objective of all rights are human rights. Human rights, however, are all geared towards promotion, protection, enhancement and maximization of human life. Kanyike states that "In traditional Africa, procreation—the reproduction and transmission of human life—is one of the most important values, if not the most important value in life.

An individual is simply not alive, if he/she is not engaged in transmitting life to another human being."[288] Thus, like many other scholars of African cultures Kanyike concludes that "Life is the greatest preoccupation of the African...Everything is centered on the communication of life, participation in that one life, its conservation and its prolongation."[289] No matter how broken human life is, it is held with almost absolute dignity and respect. The centrality of life in Ubuntu is the reason behind minority empowerment.

Due to the centrality of life in the culture of Ubuntu, marriage occupies a central place. Mbiti notes that "marriage is a duty, a requirement from the corporate society, and a rhythm of life in which everyone must participate. Otherwise, he who does not participate in it is a curse to the community, he is a rebel and a lawbreaker, he is not only abnormal but 'under human'."[290] Celibacy is inconceivable as Kanyike observes: "No one remains celibate just for the sake of it or in order to be free and no society can ever set celibacy as an ideal without running into the danger of extinction."[291] In Mbiti's interpretation celibacy is an abnormality. It is an offence against the primitive command "to increase and to multiply," and against

---

[284] Bujo (2001, pp. 3–11).

[285] Bujo (2001, p. 6).

[286] Bujo (2001, p. 6).

[287] Bujo (2001, p. 1).

[288] Edward Kanyike, *The Principle of Participation in African Cosmology and* Anthropology (Balaka: Monfort Media, 2004), p. 139.

[289] Kanyike, *The Principle of Participation in African Cosmology and* Anthropology, p. 139.

[290] Mbiti (1990, p 130).

[291] Kanyike, *The Principle of Participation in African Cosmology and Anthropology*, p. 140.

'immortality'.[292] The right to life (even for the unborn, which implies the duty to generate life) is the center of all rights. The precedence of communal life over individual life in the culture of Ubuntu is based on the logic of utilitarian maximization of the greatest good, which in the case of Ubuntu is life.

The whole community is geared towards promotion of life. If an individual proves to be an obvious impediment to the community's concern with each and all life, that individual is suppressed or eliminated. The life of the community precedes each individual life. The community is the foundation of individual life. It is the community which, not only defines and enables individuation, but individuation is absurd if not based on the community. Using the words of Benhabib Seyla, "Individuation does not precede association; rather it is the kind of associations which we inhabit that define the kinds of individuals we become."[293]

For this reason Mbiti states that in sub-Saharan Africa the "community must therefore make, create or produce the individual; for the individual depends on the corporate group…Physical birth is not enough: the child must go through rites of incorporation so that it becomes fully integrated into the entire society."[294] Consequently, the association must precede individuation. The community as a whole and the morally mature members of the community are responsible for each of its members, especially the disadvantaged and those with disabilities.

Due to the centrality of human life in the culture of Ubuntu, minority enablement and empowerment is naturally ascertained by the community in a very natural way. Decent minimum for all is ascertained in a variety of ways. Tangwa notes, for example, that "in traditional Africa practitioners of the medical and healing arts, like many other artists and specialists, normally did not charge any fees for their services" however, patients who were treated, as a matter of unspoken sense of justice and custom, "always voluntarily came back with appropriate gifts and rewards for their healer/doctor…Nso' traditional society, for instance was organized in such a way that what one needed for mere survival was at the disposal and within the reach of all and sundry."[295]

Tangwa also points out that land, being a major means of production, was not owned individually. The king ascertained that everybody who needed land for cultivation or building got it and that nobody had more than he needed. It was a taboo among the Nso' "to sale or otherwise commercialize certain things, such as the staple food, housing, water, fuel-wood, etc." Nyerere notes the same thing. He writes that in traditional African society no one was allowed to fall below the acceptable poverty line, just as no one was allowed to rise above an acceptable ceiling of richness relative to average community wealth. This spontaneous and almost natural unanimous agreement is based on the recognition of human dignity and equality, as Nyerere later observes.[296]

---

[292] See Mbiti (1970, pp. 174–175).

[293] Benhabib (1997, p. 73).

[294] Mbiti (1990, p. 106).

[295] Tangwa (2010, p. 77).

[296] Julius Nyerere, *Freedom and Socialism. Uhuru na Ujamaa* (Dar es Salaam: Oxford University Press, 1985), p. 338.

Necessities of life such as food, clothing and temporary shelter were given out or simply taken as needed.[297] Such practice would ascertain human life and dignity for all. Production of wealth in the culture of Ubuntu was never based on competition. Amassing wealth for individual security or for immediate family security only is anathema. Production was for the self without excluding the disadvantaged.[298] It is a shame for the entire society to have destitute people. It is unjust, inhuman, antisocial and an ethical/ moral immaturity on the part of the society to have desperately poor in their midst. It always meant that the society was in decadence and perishing. In sum, it is a moral obligation to help those in need. By the virtue of their being human, the poor and the people with disabilities who have a just claim to the labor, talent and time of fellow humans in whose lives they share. It is a moral duty and obligation to provide for those who cannot provide for themselves.

2.3.3.2 Minority Empowerment in Ubuntu is Based on Human Equality

Ubuntu's stance on empowerment of the minority is founded on a deep rooted understanding of human equality. It is also rooted in the fact that nobody is self-sufficient or perfect. Humans need each other. As a result of this understanding, every person in the society is equally important and a gift to every other person in it. The ability to empower the minority and 'going an extra mile' for them determines both personal and societal fulfillment and moral maturity. Personal fulfillment or actualization as human is based on the ability to engage and help other people in the community.

Using Ramose's words, "to be a human be-ing is to affirm one's humanity by recognizing the humanity of others and, on that basis establish human relations with them,"[299] whereby establishing human relations with other humans means engaging them and enabling them to the extent of their need and your ability. Consequently, "Ubuntu supports the Biblical teaching that there is more joy in giving than in receiving." (Acts 20:35). Human equality facilitates care and creates community. It can fairly be concluded that sub-Saharan indigenous Africa cannot conceive of humanity completely cut off from community.

Ubuntu's belief in minority empowerment and human equality is based on Ubuntu's communitarianism. Ubuntu communitarian world view holds that if one member of the community is suffering the whole community suffers. One cannot separate oneself from needy members of the community. Ignoring minority is a direct attack on Ubuntu communitarianism. Gyekye writes that "Communitarianism immediately sees the human person as an inherently communal being, embedded in a context of social relationships and interdependence, and never as an isolated,

---

[297] Tangwa (2010, p. 77).

[298] See Marquard and Standing (1939, pp. 20–32).

[299] Ramose (2002, p. 42).

atomic individual."[300] Bujo perceive the Bantu communitarianism as a worldview. It is not based only on humans. It involves the entire cosmos. He writes, "In the African world-view, all things hang together, all depend on each other and on the whole. This applies particularly to human beings who are closely connected with each other and with the ancestors and God."[301] Bujo further explains that this Bantu worldview which is based on Africans' experience of the world is ontological, spiritual and eschatological. He writes, "The way they think and feel is in union, not only with other people around them, but, indeed, with the deceased, even God, and the entire universe is drawn into this flow of life."[302]

All the values that increase bonding between different people within the community were considered virtue. The values that break the bond between members of the community are considered vices. That is why Broodryk writes "Ubuntu demands respect for all other human beings irrespective of race, gender, beliefs, class, and material possessions: all are equal beings reliant on each other for a happy life."[303] Equality between human beings was based on the ontological fact of being human. Everybody is recognized, given attention and engaged by everybody else. To ignore others is considered immoral since everybody commands attention of everybody else. Metz sums up this state of affairs which has been researched by many scholars into a moral principle. He states that it is immoral "to ignore others and violate communal norms, as opposed to acknowledging others, upholding tradition and partaking in rituals."[304]

Ubuntu human equality is on the basis of subsidiarity. There is a systematic spontaneous agreement that everybody should, in his or her capacity, be helped to participate in the life of the community. Production is based on ability and distribution on need. Leopold Senghor attempts to define and explain and distinguish African communitarianism, which is based on equality, participation, inclusion and sharing of life, from what he called "collectivist society" using relativistic and comparative language. He states, "The collectivist society inevitably places emphasis on the individual, on his original activity and his needs. In this respect the debate between 'to each according to his labour' and 'to each according to his needs' is significant." According to Senghor, Ubuntu is not Collectivist in approach. He states that "Negro-African society puts more stress on the group rather than on the individual, more on *solidarity* rather than on the activity and needs of the individual, more on the *communion* of persons rather than on their autonomy." However, the value of the individual along with his or her basic human rights remains indispensable. Senghor clarifies, "ours is a *community* society. This does not mean that it ignores the individual, or that collectivist society ignores solidarity, but the latter bases this

---

[300] Gyekye, "Person and Community in African Thought," in *Philosophy from Africa: A Text with Readings* ed. P. H. Coetzee and A. P. Roux (Durban: International Thomson Publishing Southern Africa, 1998), p. 319.

[301] Bujo (1992, p. 22).

[302] See Bujo (2001, p. 88); Masolo (1994, p. 498).

[303] Johann Broodryk, *Understanding South Africa—the uBuntu way of* living (Waterloo: uBuntu School of Philosophy, 2007) p. 40.

[304] Thaddeus Metz (2007, p. 237).

solidarity on the activities of individuals, whereas the community society bases it on the general activity of the group."[305]

In effect, individual contribution to the common good is not pronounced within Ubuntu culture. The maxim is "from each for all and all for each." Ruch verbalizes this mind set best when he states, "What I am myself for and by myself, matters less than what I am with, in and through the others."[306] Nyerere explains that within the culture of Ubuntu there was neither room nor tolerance for exploitation. He states, "In traditional society, everybody worked for his or her personal needs and for the needs of the extended family or ethnic group. Caring for the wellbeing of the sick, children, elderly and those with disabilities was a responsibility of each individual member of the society and of the society as a whole."[307]

Ideally, the culture of Ubuntu expects everybody to be responsible for everybody else in the community. Children, for example, belonged to the extended family and to the entire clan and tribe. Every adult would discipline or teach any child. Caring for people with disabilities is a responsibility of everybody. They need to be helped to feel equal to other members of the society. Nyerere notes that "in Ubuntu, the people with disabilities, the sick, the orphaned, widows or elderly members of the society are automatically protected so that they do not feel insecure or inferior to the rest of the members of the society." No one would be at peace if a minority is in need. The minority is a responsibility of everybody else. Any morally mature person should naturally take upon himself to address the plight of the minority in his environment. There is a delicate balance between individual property and common property. Nyerere elaborates on this fact when he writes "If a member of an ethnic group is prosperous, the whole ethnic group is prosperous. If the ethnic group is prosperous, each member considers himself or herself prosperous"

Ubuntu ascertains that everybody has the means necessary for production and that exploitation is discouraged. This was achieved as Nyerere notes by common ownership of the major means of production. "Land is communally owned in that no one has absolute right to it. Members of the community use it according to need. Laziness or refusal to work is a curse and source of shame to the respective individual and his/her family."[308] To underline African deep rooted communitarianism based on human equality Nyerere writes elsewhere that, "all basic goods were held in common, and shared among all members of the unit.

There was an acceptance that whatever one person had in the way of basic necessities, they all had; no-one could go hungry while others hoarded food." The gap between the richest and the poorest is minimized as a matter of virtuous society.

[305] Senghor (1964, pp. 93–94).

[306] E. A. Ruch, "Towards a Theory of African Knowledge," in *Philosophy in the African Context.* D. S. Georgiades and I. G. Delvare (Johannesburg: University of Witwatersrand Press, 1975), p. 18.

[307] Nyerere (1968, p. 4).

[308] Julius K. Nyerere, "Ujamaa—The Basis of African Socialism," in *I am Because We Are: Readings in Black Philosophy*, ed. Fred Lee Hord (Mzee Lasana Okpara) and Jonathan Scott Lee (Amherst, Boston: University of Massachusetts, 1995), pp. 65–72.

Nyerere observes that "within the extended family, and even within the tribe, the economic level of one person could never get too far out of proportion to the economic level of others."[309]

Ubuntu world view does not consider enabling or helping a needy person as a matter of choice or charity. One is obliged to share that which is necessary to make another human being live a dignified life. If one has more than he needs and another member of the society does not have the basic needs, the wealthy is considered as an immoral person. Refusal to provide for the basics of life is a moral omission which makes one a criminal.[310]

In sum, Bantu ethics is inseparable from human life lived in community and based on acceptance of human basic equality. Human rights in Ubuntu are rights because of the dignity of human life, its equality with any other human life and its helplessness independent of the community. It can safely be stated that the essence of Ubuntu ethics is human life in the context of community of human equals.

2.3.3.3   Minority Empowerment as a Matter of Religious and Ethical Imperative

Minority empowerment is not only an ethical imperative, it is a religious imperative. The objective of Ubuntu is tranquil and harmonious coexistence between humans and between humans and the cosmos. This objective is both ethical and religious because it supports life. The community is at the service of each life within it. God's will is order, peace and tranquility which are an optimal context for nurturing and protection of each human life. Like Mbiti, Bujo, Kasenene, Tangwa and Shutte, Onah observes that "The promotion of life is therefore the determinant principle of African traditional morality and this promotion is guaranteed only in the community." Consequently, community becomes necessary for the sake of life. The importance of community for human life is not only ethical but religious as well. Onah states that "Living harmoniously within a community is therefore a moral obligation ordained by God for the promotion of life. Religion provides the basic infra-structure on which this life-centred, community-oriented morality is based."[311]

Failure to enable and empower the minority works against the objective of Ubuntu because it violates life. Flourishing of their lives depends on those who are able in the community. Every person is religiously and ethically responsible for all life in accordance to his ability and enablement.[312] Onah concludes that "Living harmoniously within community is therefore a moral obligation ordained by God

---

[309] Julius K. Nyerere, *Freedom and Socialism. Uhuru na Ujamaa* (Dar es Salaam: Oxford University Press, 1968), p. 338.

[310] Broodryk, *Ubuntu. Life Lessons from Africa* (Pretoria: Ubuntu School of Philosophy, 2002), p. 8.

[311] Onah (2012).

[312] Bujo (2001, p. 2, 88).

for the promotion of life."[313] In line with Onah, Desmond Tutu writes, "harmony, friendliness, community are great goods. Social harmony is for us the *summum bonum*—the greatest good.

Anything that subverts or undermines this sought-after good is to be avoided like a plague. Anger, resentment, lust for revenge, even success through aggressive competitiveness, are corrosive of this good."[314] Failing to pay attention to, and address the plight of the minority is considered a violation of harmonious community life. One is not only guilty before oneself and the community for failing to empower the minority; he or she is responsible and culpable before God for the omission.

Minority empowerment among the Chagga people of Kilimanjaro Tanzania is much more sophisticated and realistic. However, it is one of the best examples of Ubuntu as practiced in real life with regards to minority empowerment. For the Chagga people instead of giving a poor person milk the poor person is helped to own a cow. However he has to prove over time to the society that he can assume the responsibility of taking care of the cow. He doesn't get to own it instantly.

He keeps the cow as borrowed property, gives back to the owner the first calf produced by the cow, then own the second calf; then the cycle repeats itself until he or the owner decides to terminate the contract. The Chagga of Uru calls this practice *iarà* (*iarà* is infinitive which means lending with an intention to help another person help himself. The verb and root of *Iarà* is *arà*). *Iarà* redeems the poor person from his misery, enabling him to salvage himself, be independent and be responsible. Interestingly, this practice is an application of the principle of subsidiary and a perfect illustration of recognition of human equality. *Iarà* is enablement per excellence. *Iarà* is an illustration of not only the presence of ethical principles within indigenous Bantu people but of a highly developed practical ethics, concept of justice, fairness, responsibility and human equality.

Minority empowerment is necessary for a peaceful community. For sub-Saharan Africans peace is not merely an absence of war and active conflict. Rather peace is conceived "in relation to order, harmony, and equilibrium." Peace in the universe is not only ideal for the survival of human life and other lives, but the will of God. God wills that there is harmony and favorable equilibrium in the universe. "The order, harmony and equilibrium in the universe and society is believed to be divinely established and the obligation to maintain them is religious." Peace is a moral value because its attainment and sustenance requires human proactive and initiative participation.[315]

Sub-Saharan Africans believe that the order ordained by God is upset when any human life is not treated in accordance with its due dignity and respect. The order is upset when there is no ontological, religious, social and economic equality among human beings. For both human dignity of the minority and equality of humanity, minority empowerment is *sine qua non*. If the minority is not empowered there can be no peace within the majority or the minority.

---

[313] Onah (2012).
[314] Tutu (1999, p. 35).
[315] Onah (2012).

"Peace is good relationship well lived; health, absence of pressure and conflict, being strong and prosperous…"[316] "Peace is the totality of well-being: fullness of life here and hereafter…'the sum total of all that man may desire: an undisturbed harmonious life.'"[317] Absence of peace means, at the same time, a moral evil. According to Bujo personal health is contingent to community and the cosmos. Bujo concludes, "Health, therefore, implies safe integration into the bi-dimensional community as the place where life grows."[318] This means that personal health cannot exclude the minority in the community.

The ideal of health is on-going growth into bonding with other humans, especially by addressing recognizing their humanity, engaging it as an equal partner. In Broodryk words, it "is to become more fully human which implies entering more and more deeply into community with others."[319] Life as such is not completely a personal concern. To a very large extent all life belongs to the immortal community. The individual is "just a link in the chain uniting the present and future generations,"[320] using the words of Kasanene. It is the concern of everybody to bring every life to its fullness to the best of his ability.

Desmond Tutu explains the ideal personal stance towards other people from Ubuntu perspective in these words, human beings "are diminished when others are humiliated, diminished when others are oppressed, diminished when others are treated as if they were less than who they are."[321] In other words, failure to empower minority in the society is not only a violation against them, it violates also the humanity of the subject who ignores the minority.

The community expects everybody to engage and empower the minority as a way of affirming not only the humanity of the minority but, especially, his own humanity.[322] Among the Chagga people of Tanzania, if one harvests crop from his land, he or she should leave a little portion on the land for the needy. The minority naturally know that it is meant for them. Among most Bantu people who are travelling don't carry much food with them. They would stop at any community village on their way and expect to be given something to eat, a drink and a place to spend the night if tired.

Ubuntu stance towards the minority is in line with what John Finnis recommends in his work *Natural Law and Natural Rights*. Nature of property rights requires it.[323] Julius Nyerere points out that in the traditional society the minority were protected so that they did not feel insecure or inferior to the rest of the members of the society.

---

[316] Robert Rweyemamu, "Religion and Peace," p. 381.

[317] J. S. Awolalu, *The Yoruba Philosophy of Life*, in Robert Rweyemamu, Religion and Peace," p. 382.

[318] Bénézet Bujo (1998, p. 182).

[319] Broodryk, *Ubuntu. Life Lessons from Africa*, p. 101.

[320] Kasanene (1994, p. 349).

[321] Tutu, "Ubuntu and Indigenous Restorative Justice." http://www.africaworkinggroup.org/files/UbuntuBriefing3.pdf. February 15, 2012.

[322] Broodryk, *Ubuntu: Life Lessons from Africa*, p. 8.

[323] Finnis (1980, pp. 186–187).

From the perspective of Ubuntu culture prosperity of one member of the community was considered prosperity of the whole community.[324] As a way of assuring the decent minimum for all, and equality of access and ownership of the major means of production land and other major means of production is basically communally owned in that, no one has absolute right to it. This mode of owning and using major means of production ascertained inescapability of communitarianism and assurance of enablement and subsidiarity for all. Community members use it according to need and ability for self and the society.[325]

One ought to work for oneself and for the minority. Refusal to work is equivalent to suicide because it implies cutting oneself from the community.[326] Consequently Broodryk observes that caring for oneself and for other members of the community through human labor is a moral imperative in Ubuntu. Thus, responsible "Caring is an important pillar in the *Ubuntu* worldview."[327] Since care enables one to realize his humanity, Michael Battle argues that the minority helps the majority to realize their humanness in the very act of recognizing and empowering the minority.[328] Thus, Mnyaka and Motlhabi are justified when they state that "Ubuntu ethics is anti-egoistic, as it discourages people from seeking their own good without regard for, or to the detriment of, other persons in the community."[329]

Minority empowerment is within the kernel of Ubuntu worldview. It is ethical, social, religious and psychological imperative. Deliberate refusal to engage and empower the minority is self-defeating since it means annihilating one's own humanity by estranging him or her from oneself, from the community and from God.

2.4 Conclusion

In Ubuntu ethics, the community determines and defines individual rights and obligations. Even though individuals have innate individual dignity, Ubuntu assumes that the welfare of individuals is dependent on the welfare of the community as a whole, just as it assumes that 'being an individual is being with others' and that the self stands in constant need of an-other. Consequently the community takes precedence over its constituent individuals. Even though Ubuntu ethics recognizes the individual's need for the community for survival, self-definition, development and actualization, every individual remains unique and with autonomy.

Since each person has a right to self-determination, there is inevitable tension between individual rights and universal rights. Individual rights being subordinate to universal rights, there cannot be absolute individual rights in Ubuntu. This tension,

---

[324] Nyerere, "Ujamaa—The Basis of African Socialism," pp. 65–72.

[325] Nyerere, "Ujamaa—The Basis of African Socialism," pp. 65–72.

[326] Nyerere, "Ujamaa—The Basis of African Socialism," pp. 65–72.

[327] Broodryk, *Ubuntu. Life Lessons from Africa*, p. 48.

[328] Battle (1997, p. 65).

[329] Mnyaka and Motlhabi (2003, p. 224).

however, is inevitable since existence itself is a web of interconnections, interactions, and symbioses between humans and between humans and the non-human part of the universe.

The tension between individuals and the community in Ubuntu ethics is managed by an on-going process of initiation into the wider community. Initiations are geared toward acknowledgement that ethically, individual rights meet their limit in the rights of other individuals represented in sum by the community. It is the continual process of initiation which enables sub-Sahara Africans to think in 'both/ and rather than either/or' categories. In other words, individual autonomy is not practicable if it doesn't recognize other persons' right to autonomy. The community ascertains that. Since individuals realize their humanity in their relationships with other humans, the tension between individual rights and universal rights is constructive as it enables and facilitates cognitive and moral development.

From the perspective of Ubuntu, the poor and the underprivileged have a just claim to the labor, talent and time of the community in whose life they share. It is a moral duty to provide for those who cannot provide for themselves while recognizing and appreciating their contribution, according to the principle of subsidiarity. No human life is in vain. When human life is at stake, no individual rights holds. Human life overrides all individual rights, except when such life is a threat to more lives or the life of the community.

Ubuntu ethics not only recognizes cognitive and moral development with regards to ethical maturity, which in Ubuntu is equivalent to the ability to care, it facilitates the process. When an individual has objectively been proven to be mature, such individual is allowed to transcend the limitations and boundaries imposed by the community and act freely. Such individuals are allowed to do so because they are believed to be really mature, which means they always act in the interests of the community as they act in their individual interests. Recognizing human dependence on the biosphere and the cosmos, Ubuntu recognizes non-human biospheric and cosmic rights. Humans have duties and obligations to provide good stewardship, treasure and safeguard their environment for the current and for future generations as a matter of ethics.

Having analyzed the components of Ubuntu, clearly, at the core of Ubuntu is ethics of care. The following chapter explores ethics of care as it enlightens Ubuntu and as it is enlightened by Ubuntu. Ethics of care recognizes individual rights having merits because they have universal meaning. Individual and universal rights need to be interpreted in light of ethical responsibility having meaning within human relationships. There is need for reciprocity of care that clarifies the meaning of ethical responsibility.

Chapter 3
Ethics of Care: Enlightening the Role of Rights in Global Bioethics

There has been some opposition concerning acceptability of care as an ethic. All-mark, for example, argues that, in itself, care is "morally neutral," except when "it is for the right things and expressed in the right way." Thus, in his own words, "'Caring' ethics assumes wrongly that caring is good … 'caring' ethicists take the fact that care-related terms are used to express moral judgment to imply that care is itself a good, or the good. This inference is both invalid and false."[1] According to Allmark, therefore, the whole concept of care ethics is fallacious and empty of substance. He argues, "*mutatis mutandis,* a caring person, is not someone who cares indiscriminately. She is someone who cares in the core sense about the things she ought to care about, and to the right degree." Only then such a person can claim that her care is morally evaluable and justifiable as good or bad. In sum, Allmark contends that "focusing on care as a moral quality in itself, something it is not, the ethics of care can tell us nothing of what those right things (the objects of care) are."[2] However, this book assumes acceptability of care as a valid ethic which has been globally recognized and which cannot be ignored or sidelined.

According to Ubuntu philosophy, care is not only an ethic; it is the *conditio sine qua non* for the possibility of genuine ethics. In other words, ethics is based on the human ability and essential characteristic to care. All principles of ethics are based on the fact that human beings are caring creatures. Care is assumed and presupposed in human interactions. All principles of ethics are derived from, and aim at care. In my view, therefore, care is not only one of the many ethics. Care precedes ethics. It is that for which ethics exist. Defending indifferent or uncaring ethics is *reductio ad absurdum* as it renders ethics purposeless and meaningless. I contend that care should neither be considered feminine ethics nor be viewed as one of the many kinds of ethics. Care transcends those categorizations. Ubuntu philosophy is about care for humans and the universe.

---

[1] Peter Allmark Sheffield and North Trent College of Nursing and Midwifery, Sheffield. "Can there be an ethics of care?" *Journal of Medical Ethics* 21 (1995), 19.

[2] Peter Allmark Sheffield and North Trent College of Nursing and Midwifery, Sheffield. "Can there be an ethics of care?" 23. The words in the brackets are mine.

L. T. Chuwa, *African Indigenous Ethics in Global Bioethics,* Advancing Global Bioethics,   89
DOI 10.1007/978-94-017-8625-6_3, © Springer Science+Business Media Dordrecht 2014

Ubuntu recognizes the tension between individual and universal rights. The meaning of this tension can be enlightened by considering the Ethics of Care. The first major component of Ethics of Care concerns individual rights having merit because they have universal meaning. The second major component of Ethics of Care concerns human relationships. Individual and universal rights need to be interpreted in light of ethical responsibility having meaning within human relationships. The third major component of Ethics of Care concerns reciprocity of care. To integrate the debate on individual/universal rights and relationships, there needs to be reciprocity of care that clarifies the meaning of ethical responsibility. This section explores competing individual rights in relation to moral development; human relationships with regards to morality in general and narrative in particular; and reciprocity of care with special emphasis on the role of context in ethics and the problem of universalization of care.

3.1   Individual Rights

Rather than contradict the idea of care, individual rights facilitate organized care. Ubuntu neither contradicts the idea of individual human rights nor that of care ethics. Ubuntu bridges them. The first major component of Ethics of Care concerns individual rights. Individual rights have merit because they have universal meaning. This component has two related concepts. The first concept concerns rights and moral development as seen from the perspective of Lawrence Kohlberg. The second concept concerns competing rights contrasted with care and responsibility as seen from the perspective of both Lawrence Kohlberg and Carol Gilligan.

3.1.1   Rights and Moral Development

The principle of moral development is based on, and concerned with the conception of justice.[3] Kohlberg's major assumption is that a human being's conception of justice develops in stages. Kohlberg's premise is based on Jean Piaget's psychological theory of human development. According to Kohlberg, the process of moral development is concerned with the conception of justice. His focus is on how people justify behaviors.[4] Kohlberg proposed six stages of moral development, based primarily on age. The six stages can be grouped into three levels: Pre-conventional (hedonistic stage: physical consequences of human action determine their moral value), conventional (conformity with authority and society/meeting expectations of others determine morality of human action) and post-conventional (discernment of moral value of human action should be independent of their physical rewards, societal expectations, authority and personal bias). The pre-conventional stage includes two stages: first, obedience and punishment; second, an orientation of self-interest.

---

[3] Kohlberg (1981).

[4] Kohlberg (1958).

The conventional level includes two stages: first, interpersonal accord and conformity; second, authority and social-order maintaining orientation. The post-conventional level has two stages: first, social contract orientation; second, universal ethical principles.[5] The main implication of Kohlberg's theory of moral development is that since fairness depends on how one perceives justice, and perception of justice is relative to the six stages; ethical fairness for each human person is relative to the six stages of development. However, according to Ubuntu perception and pursuit of justice in itself should be based on and motivated by human nature and desire to care.

Kohlberg noted that few people get to the fifth stage (social contract orientation) and even fewer get to the sixth stage of moral development (universal ethical principles). That being the case, actual maturity in universal ethical principles would be unattainable by the majority of the human population. However, moral development is undeniable. Kohlberg's sixth stage of moral development (the universal moral principles) is, actually the objective of all ethics and it is, from Ubuntu perspective, care actualized. According to Ubuntu a really mature person does not need any regulations or principles as they have grown to attain the objective of ethical principles, which is authentic care. Actions of such a person are caring actions as they are both motivated by and seek care for all.

3.1.1.1 A Case Against *Bag of virtues*/Indoctrination Approach to Morality

Kohlberg criticizes and refutes the traditional theory of moral development, the so called 'bag of virtues' approach to morality. The traditional theory, which has been accepted universally, assumes that "moral values are not universal, that they are culture relative, and that they are not innate." If the traditional theory of moral development is accepted as valid, a child would be morally a mere potency that is helplessly and totally dependent on the society for any moral development since potency has to be given a form to be real. Such assumption denies a child not only responsibility for its moral development; it denies the child its innate human nature making the society totally responsible for the moral development of each of its members. In a similar manner Kohlberg argues against relativity of moral principles. He contends that "there are in fact universal human ethical values and principles." And such values are not infused by the community, they essentially develop from within.[6]

According to Kohlberg the traditional approach to moral development is essentially "indoctrination of conventional or social consensus morality." In his view the relativistic theory of morality leads to absurdity. In sum it is destructive of actual personal moral formation and maturation. In Kohlberg's own words relativistic theory of morality "is a theory of virtue that commends itself to the 'commonsense' of those whose view of morality is conventional." Kohlberg disqualifies such an approach because it is based on "social relativism, the doctrine that, given the relativity

---

[5] Kohlberg (1971, 1983); Colby et al. (1983).

[6] Kohlberg (1971, p. 32).

of values, the only objective framework for studying values is relative to the majority values of the group or society in question." It lacks authenticity and credibility.[7]

Kohlberg observes that human beings share the same basic universal moral values regardless of social consensus. Different decisions that particular persons make, whether morally correct or erroneous, do not change the universal innate tendency toward moral goodness. Social experience may be helpful but it is not the source of morality. Kohlberg posits that "our values tend to originate inside ourselves as we process our social experience." Consequently, human race share the "same basic moral values." Culturally specific practices differ greatly but the principles underlying such practices or beliefs are constant and common to all.

The difference in specifics differs due to social experience and environment (such as: eating squirrels is wrong, sharing a room with one's mother-in-law is wrong) do not "engender different basic moral principles (for example, consider the welfare of others, treat other people equally)." Kohlberg attributes difference in basic moral values to "different levels of maturity in thinking about basic moral and social issues and concepts." However, he realistically acknowledges the role of society in individual moral maturity. He states that "exposure to others, more mature than ourselves helps stimulate maturity in our own value process."[8] To support his argument about universality and innate nature of moral values Kohlberg makes a presumptive statement that "All parents know that the basic values of their children do not come from the outside, from the patents, although many wish they did." To illustrate this general statement Kohlberg uses an example of his son who at the age of four joined a pacifist movement and became vegetarian in protest to killing of animals for meat. Listening to the story of Eskimo seal hunting, the same child remarks, "'you know, there is one kind of meat I would eat, Eskimo meat. It's bad to kill animals so it's all right to eat Eskimos [because they kill animals]'." According to Kohlberg this simple observation of his own child's moral reasoning makes clear two important points: "(1) that children often generate their own moral values and maintain them in the face of cultural training, and (2) that these values have universal roots."[9]

Kohlberg fails to see the connection between innate moral principles and the innate human nature to care. Ubuntu does. Humans care about themselves, so they self-preserve. Ubuntu recognize that self-preservation that is extended to other individuals is indicative of moral maturity. A person is mature to the degree he can care for his neighboring persons and environment. Personal maturity increases with the personal ability to universalize care. According to Ubuntu, a fully mature person can sacrifice his life for the good of others and the world if situation calls for it.

3.1.1.2 Kohlberg's Explanation of Dynamics Behind Human Development

One of the most basic assumptions of the developmental theories noted by Kohlberg is that "development involves basic transformations of cognitive structure which

---

[7] Kohlberg (1981, p. 2).

[8] Kohlberg (1981, p. 14).

[9] Kohlberg (1981, pp. 14–15). The words in the brackets are mine.

cannot be defined or explained by the parameters of associanistic learning." Notably, this assumption rules out the traditional theory of development which explains Psychosocial and moral development in terms of human interactions and relationships exclusively. According to this new understanding, development "must be explained by parameters of organizational wholes or systems of internal relations."[10]

There is real relationship between cognitive development and both the biosphere and the cosmos. Kohlberg notes that development of cognitive structure is induced by, and results from "processes of interaction between the structure of the organism and the structure of the environment."[11] According to Kohlberg, proper cognitive development seeks reconciliation, harmony, balance and healthy equilibrium between an individual and the environment. Greater equilibrium between an organism and its environment is the ideal of cognitive development. This means "greater balance or reciprocity between the action of the organism upon the (perceived) object (or situation) and the action of the (perceived) object upon the organism." Even though optimal equilibrium is hardly measurable, leave alone being attainable, it always remains the ethical ideal; not only for human individuals but also for the biosphere and the cosmos.[12] Cognitive development requires not only the subject but also the object since it is by nature interactive and reciprocal. Kohlberg posits that "Cognitive structures are always structures (schemata) of action. While cognitive activities move from the sensorimotor to the symbolic to verbal-propositional modes, the organization of these modes is always an organization of actions upon objects."[13]

Kohlberg underlines the role of unity in personal development. Development may be termed physical, social, psychological, emotional or moral but it all refers to the same ego or self. He states that "these strands are united by their common reference to a single concept of self in a single social world." Selfhood as a unity is essential in personal development. The self is not only the nucleus; it is both the basis and target of personal development. Social relationship is secondary to selfhood. Kohlberg writes that "Social development is, in essence, the restructuring of the (1) concept of self, (2) in its relationship to concepts of other people, (3) conceived as being in common social world with social standards."[14]

Thus, Kohlberg's observation of the importance of the self as unity does not undermine or downplay the role of other selves and the cosmos in the conception and development of the self. Different aspects of the self are equally important for self-development. In Kohlberg's words, "there is no distinction between the affective and the cognitive in terms of precedence or importance". They both are equally important representing "different perspectives and contexts in defining structural change."[15] Even though physical and cognitive development is not based entirely on social interactions, Kohlberg notes that "All the basic processes involved in 'physical'

---

[10] Kohlberg (1969, p. 348).
[11] Kohlberg (1969, p. 348).
[12] Kohlberg (1969, p. 348).
[13] Kohlberg (1969, p. 348).
[14] Kohlberg (1969, p. 349).
[15] Kohlberg (1969, p. 349).

cognitions, and in stimulating developmental changes in these cognitions, are also basic to social development." Social and moral development however, is based on role-taking. It rests on the "awareness that the other is in some way like the self." Reciprocity is an important part of social cognition and development. Social development is absurd and self-defeating if it is not based on the fact that the other is able to recognize the self, know him, relate with him and respond to the self. Reciprocity is the core of mutual complementarity and fulfillment. Kohlberg writes, "Accordingly developmental changes in the social self-reflect parallel changes in conceptions of the social world."[16]

Social and moral development tends towards optimal equilibrium that promotes harmony without undermining the ego/self. There is an ongoing balancing between "actions of the self and those of others toward the self." According to Kohlberg such social equilibrium viewed from its general perspective is the "end point or definer of morality, conceived as principles of justice, i.e., reciprocity or equality." From more personal individual perspective such equilibrium "defines relationships of 'love,' i.e., of mutuality and reciprocal intimacy." However through role assumptions and transformations in the process of social self-development, there is an inner undeniable instinct to preserve and maintain self/ego-identity in spite of inevitable adjustments.[17]

Ubuntu philosophy is in full agreement with Kohlberg regarding the contingency of personal moral development on interaction with his environment. Ubuntu realizes that a lot that shapes personal identity comes from personal interaction, engagement and balancing with one's environment. One's environment is the undeniable "other" without which there could not be the "self." From Ubuntu perspective, therefore, caring for self cannot be separated from caring for the "other."

3.1.1.3 Facilitation/Stewardship of Moral Development

According to Kohlberg, ethical and ideal cognitive/moral development should not be mere traditional "value clarification." Kohlberg's theory of moral development in particular and cognitive development in general is deductive rather than inductive. According to this deductive theory, the best way to educate is to help students to find reasons and explanations from within, making education an introspective process. Depending on the student's moral/cognitive maturity, some reasons may be a better interpretation of the universal constants than others. Although this theory can be traced all the way to Plato and Socrates, it was immediately borrowed from Blatt (Kohlberg acknowledges it). The theory is based on the assumption that there are innate universal goals and principles. Such universals are neither culture-specific nor relativistic in nature. They are constants which go deeper than the changing cultural values.[18]

---

[16] Kohlberg (1969, p. 349).

[17] Kohlberg (1969, p. 349).

[18] Kohlberg (1981, pp. 27–28).

The main difference between Kohlberg's proposed methodology (which I refer to as stewardship or facilitation) and the traditional methodology is that while the traditional method of moral and cognitive formation is indoctrinative, consequently patronizing, by "moving the student in the direction of accepting the teacher's moral assumptions," Kohlberg's approach "avoids preaching or didacticism linked to the teacher's authority."[19]

Traditional moral education "reflects the unconscious wisdom of society and its needs for 'socializing' the child for his own welfare as well as that of society." While such approach is not ill intentioned, it may not be beneficial for the recipient of education. It may actually deny the recipient an opportunity to develop authentically. Kohlberg states that "when such 'socialization' or rule enforcement is viewed as implying explicit positive educational goals, it generates a philosophy of moral education in which loyalty to the school and its rules is consciously cultivated as a matter of breeding loyalty to society and its rules."[20] Since personal moral development is not mere loyalty to the society, such method of education is patronizing. Assuming that a child acquires moral values and principles by internalizing cultural norms reduces development into "direct internalization of external cultural norms." Personal development is thus reduced to conformism. "The growing child is trained to behave in such a way that he conforms to societal rules and values."[21]

Just as conformism is detrimental to cognitive and moral development, so is value relativism. Kohlberg describes value relativism as "both a doctrine that 'everyone has their own values,' that all men do not adhere to some set of universal standards, and a doctrine that 'everyone ought to have their own values,' that there are no universal standards to which all men ought to adhere." Logically, value-relativity leads to irrelevance of ethics and morality since it relativizes ethical principles and values. In Kohlberg's words, "value-relativity position often rests on logical confusion between matter of fact, what 'is,' and matter of value, what 'ought to be.'"[22]

Real developmental moral education is "neither an indoctrinative nor relativistic classroom discussion process."[23] Since there are universal objective moral principles which "transcend both individual personal differences and cultural specific differences," indoctrination may be harmful to cognitive and moral development because "moral development is directly related to cognitive development."[24] According to Kohlberg authentic ethical resolution of moral problems can be achieved by "creating a democratic classroom in which issues of fairness are settled by discussion and a democratic vote."[25] Although through this method, objective universality is compromised by the moral and cognitive maturity of participants, the approach

---

[19] Kohlberg (1981, p. 28).

[20] Kohlberg (1971, p. 30).

[21] Kohlberg (1969, p. 30).

[22] Kohlberg (1969, p. 33).

[23] Kohlberg (1981, p. 28).

[24] Kohlberg (1971, pp. 40–45).

[25] Kohlberg (1981, p. 28).

is more ethical as it respects each participant's autonomy and empowers their authentic development.

Kohlberg contends that moral thought "seems to behave like all other kinds of thought. Progress through the moral levels and stages is characterized by increasing differentiation and increasing integration, and hence is the same kind of progress that scientific theory presents."[26] Kohlberg refutes the traditional assumption that "morality and moral learning are fundamentally emotional, irrational processes."[27] On the contrary, moral education permeates all aspects of human life. It requires "multi-disciplinary approach. It requires sociological and psychological approach. Moral education cannot ignore social psychology."[28] Kohlberg notes that schools and teachers are, unfortunately, "engaged in moral education without explicitly and philosophically discussing or formulating its goals and methods."[29] This purposeless and often unintended education may be counterproductive, impeditive or even destructive of real cognitive and moral development needed. In sum Kohlberg concludes that:

1. The current prevalent definition of the aims of education, in terms of academic achievement supplemented by concern for mental health, cannot be justified empirically or logically.
2. The overwhelming emphasis of educational psychology on methods of instruction and tests and measurements that presuppose a 'value-neutral' psychology is misplaced.
3. An alternative notion that the aim of the schools should be the stimulation of human development is a scientifically, ethically and practically viable conception that provides the framework for a new kind of educational psychology.[30]

Ubuntu is partially in agreement regarding introspective method of learning and moral development. Ubuntu believes that humans have potential to learn and develop. The potential, however, remains just that—a potential, if not stimulated, and brought into actualization by one's external environment, the "other." In other words, the inter-dependence between the self and the other is indispensable both in education and in moral development. Equally, Ubuntu believes in universal values and ideals. However, universal values and ideals have to be concretized in actual life, at least partially, to be relevant for humanity. Ideals have to be engaged and in the process of actualization.

---

[26] Kohlberg (1981, p. 26).

[27] Kohlberg (1971, p. 32).

[28] Kohlberg (1971, p. 24).

[29] Kohlberg (1971, p. 29).

[30] Kohlberg (1981, p. 50).

3.1.2   Competing Rights

Kohlberg notes that moral problems arise from competing rights; their resolution depends on a proper conception of human rights.[31] Kohlberg states that the first sense of the word *moral* corresponds to a perspective that emphasizes impartiality, universalizability and the struggle to come to consensus.[32] Competing rights are reflected in Kohlberg's stages of moral development at stage 5. It is the first stage of the post-conventional level. At this stage the individual realizes that each person is a separate entity within the society and that each individual's views may take precedence over the society's views. Each tends to develop a set of principles about what is right and wrong. At this stage, rules help maintain order but even rules are subject to criticism and change since they are regarded as social contracts. Human rights and the utilitarian principle of "the greatest good for the greatest number of people" play a big role in maintaining peace and order.[33]

3.1.2.1   Six-Staged Psychosocial Moral Development Theory

Inspired by Piaget's study of structural moral development, Kohlberg develops what he describes as a "typological scheme describing general structures and forms of moral thought which can be defined independently of the specific content of particular moral decisions or actions."[34] Such structures and forms are developmental in nature, in the sense that they are stages that lead to moral maturity. Kohlberg's basic argument is that there is "definite and universal levels of development in moral thought."[35] Such stages or levels represent "separate moral philosophies, distinct views of the socio-moral world." People in one level have an objective moral perspective that is limited to that particular level or stage of moral development. Kohlberg clarifies this point when he writes "We can speak of the child as having his own morality or series of moralities."[36] Kohlberg's study reveals that children don't receive their morality from outside. It essentially comes from within them regardless their parents, or their human environment.[37]

Kohlberg's six stages are based on human life in its natural development. They are universal, in the sense that, they are not limited by socio-cultural confines. They start from human need for self-preservation in the first stage, which all children have. They then develop into the sixth stage, that of objectivity and recognition of the sacredness of human life, breaching which is self-condemnatory. Thus, the six stages are found in all cultures. Kohlberg observes that

---

[31] Kohlberg (1971, p. 51).

[32] Kohlberg, Lawrence. "The current formulation of the theory" cited by Tronto (1993, p. 87).

[33] Kohlberg and Lickona (1976).

[34] Kohlberg (1968a, p. 25).

[35] Kohlberg (1968a, p. 28).

[36] Kohlberg (1968a, p. 25).

[37] Kohlberg (1968a, p. 26).

> The social worlds of all people seem to contain the same basic structures. All the societies we have studied have the same basic institutions—family, economy, law government. In addition, however, all societies are alike because they are societies of—systems of defined complementary roles. In order to play a social role in the family, school, or society, children must implicitly take the role of others toward themselves and toward others in the group. These role-taking tendencies form the basis of all social institutions. They represent various patternings of shared or complementary expectations."[38]

In order to be able to evaluate moral development in other cultures, however, there is need to eliminate all bias. When Kohlberg decided to locate moral development in other cultures, he was advised by anthropologists to "throw away" his "culture-bound moral concepts and stories and start from scratch learning a whole new set of values for each new culture."[39] Kohlberg contends that strictly speaking, "cultural relativity of ethics, on which almost all contemporary social scientific theorizing about morality is based, is in error."[40] Hence, morality is universal to human species.

Kohlberg's main argument is that there definitely is moral development. Secondly, moral development is invariant and sequential. He states that "'True' stages come one at a time and always in the same order…In our stages, all movement is forward in sequence and does not skip steps."[41] The development is in form of differentiation, integration and universalization. In Kohlberg's own words, "each step of development then is a better cognitive organization than the one before it, one which takes account of everything present in the previous stage, but making new distinctions and organizing them into a more comprehensive or more equilibrated structure."[42] Apparently there is no regress into previous stages. Backward movement is not desirable, forward movement is. "The child in the third stage tends to move toward or into stage 4, while the stage-4 child understands but does not accept the arguments of the stage-3 child."[43]

At level one of Kohlberg's theory of cognitive moral development (pre-conventional level) "value resides in external quasi-physical happenings…physical needs rather than in persons and standards."[44] Consequently, at stage one; obedience is based on avoidance of punishment to the self. At this stage "The value of human life is confused with the value of physical objects and is based on the social status or physical attributes of the possessor."[45] At stage two of the first level "the value of human life is seen as instrumental to the satisfaction of the needs of its possessor or of other people."[46] At this stage people are still highly egotistic. The motive behind conforming is obtaining rewards.

---

[38] Kohlberg (1981, p. 26).
[39] Kohlberg (1968a, p. 29).
[40] Kohlberg (1981, p. 105).
[41] Kohlberg (1981, p. 20).
[42] Kohlberg (1968a, p. 30).
[43] Kohlberg (1968a, p. 30).
[44] Kohlberg (1958, p. 343).
[45] Kohlberg (1981, p. 19–20).
[46] Kohlberg (1981, p. 19–20).

At level two of Kohlberg's theory of cognitive moral development (conventional level) "moral value resides in performing good or right roles, in maintaining the conventional order and the expectancies of others."[47] At stage one of level two there is clear desire for approval and pleasing others. Acceptance of natural role and judgments are based on intentions. At this stage there is acceptance of duty, although it is often confused with self-interest and the assumption that authority represents moral rightness. A person at this stage would seem to say "I have deprived myself of something by conforming or working and the social order should see to it that I have not been deprived in vain."[48]

The motive behind conforming, however, is avoidance of disapproval by others. Hence "the value of human life is based on the empathy and affection of family members and others toward its possessor."[49] At stage two of level two moral values resides in blind obedience and compliance to social order for its own sake as though it is the ideal of a moral life. People at this stage have special respect for duty and social expectations. The motive behind conforming is avoidance of "censure by legitimate authorities and resultant guilt…Life is conceived as sacred in terms of its place in a categorical moral and religious order of rights and duties."[50]

At level three of Kohlberg's theory of cognitive moral development (post conventional level) "moral value resides in conformity by the self to shared or shareable standards, rights or duties."[51] At stage one of level three there is clear recognition of arbitrariness, need for rules and need for agreement. Hence duty is "defined in terms of contract, general avoidance of violation of the will or rights of others, and majority will and welfare." At this stage "Life is valued both in terms of its relation to community welfare and in terms of life being a universal human right."[52] At stage two of level three there is clear transcendence of recognized social rules to "principles of choice" based on "logical universality and consistency" the orientation tends toward "conscience as a directing agent and to mutual respect and trust." The motive behind conforming is avoidance of "self-condemnation…Human life is sacred—a universal human value of respect for individual."[53]

To explicate the different stages of moral and cognitive development Kohlberg used the famous Heinz dilemma. Basically, the dilemma is a practical moral problem resolution of which indicates one's location in Kohlberg's six stages of moral development. In Kohlberg's view the stages are universal and they "lead toward an increased *morality* of value judgment, where morality is considered as a form of judging as it has been in philosophic tradition running from analyses of Kant to those of modern analytic or 'ordinary language' philosophers."[54] Persons in

---

[47] Kohlberg (1958, p. 343).

[48] Kohlberg (1958, p. 252).

[49] Kohlberg (1981, p. 19–20).

[50] Kohlberg (1981, p. 19–20).

[51] Kohlberg (1958, p. 343).

[52] Kohlberg (1981, p. 19–20).

[53] Kohlberg (1981, p. 19–20).

[54] Kohlberg (1968a, p. 29).

different stages of moral development have different moral perspective, reasoning and judgment. Heinz dilemma helps demonstrate the difference:

> In Europe a woman was near death from a special kind of cancer. There was one drug that the doctors thought might save her. It was a form of radium that a druggist in the same town had recently discovered. The drug was expensive to make, but the druggist was charging ten times what the drug cost him to produce. He paid $200 for the radium and charged $2,000 for a small dose of the drug. The sick woman's husband, Heinz, went to everyone he knew to borrow the money, but he could only get together about $1,000 which is half of what it cost. He told the druggist that his wife was dying and asked him to sell it cheaper or let him pay later. But the druggist said: "No, I discovered the drug and I'm going to make money from it." So Heinz got desperate and broke into the man's store to steal the drug for his wife.
> Should the husband have done that? Was it right or wrong? Is your decision that it is right (or wrong) objectively right, is it morally universal, or is it your personal opinion?[55]

The post conventional level is considered the ideal of moral maturity. According to Kohlberg the post conventional levels is characterized by "a major thrust toward autonomous moral principles which have validity and application apart from authority of the groups or persons who hold them and apart from the individual's identification with those persons or groups."[56] The post conventional level of moral and cognitive development transcends cultural limitations of morality. A few people are universally known to have attained the post-conventional level of moral and cognitive maturity. In the words of Kohlberg, "Socrates, Lincoln, Thoreau and Martin Luther King tend to speak without confusion of tongues, as it were." Kohlberg attributes their moral and cognitive maturity to the fact that "the ideal principles of any social structure are basically alike, if only because there simply aren't that many principles which are articulate, comprehensive and integrated enough to be satisfying to the human intellect."[57]

Ubuntu differs substantially with Kohlberg's whole idea of moral development based on 'competing rights.' According to Ubuntu, moral maturity is not attained through competition. On the Contrary, moral maturity happens through continuous initiation and orientation into the world of the "other." It is about how one reaches out to others, considers them and cares for them. Kohlberg's post conventional stage in which an individual is considered fully mature when one acts autonomously for the course of what is right independent of validation given him by authority or society is evident in Ubuntu, except the path to that moral maturity is radically different. Moral maturity in Ubuntu is achieved by embracing others, by relating with others, by reaching out to others, by recognizing others and by a continuous process of caring reconciliation and initiation into the society and the world. Ubuntu moral stance in morality and moral maturity is that of 'one-with, rather than one-against' the community.

---

[55] Kohlberg (1981, p. 12).

[56] Kohlberg (1968a, p. 26).

[57] Kohlberg (1968a, p. 30).

3.1.2.2 Objectives and Methods of Education Should Not Contradict the Process of Moral and Cognitive Development

The cognitive development theory of Kohlberg calls to question both the content and method of education. Kohlberg questions the general objective of education from the perspective of his theory of moral and cognitive development. In his research, Kohlberg identifies three different understandings and consequential development of education ideology in the western world. The first of the three understandings is Romanticism. Originating from Rousseau, Romanticism holds that "what comes from within the child is the most important aspect of development; therefore, the pedagogical environment should be permissive enough to allow the inner 'good' to unfold and the inner 'bad' to come under control."[58] This understanding believes in the innate nature of cognitive development. It basically affirms the presence of the potential which only needs right environment to develop on its own.

The second understanding of development or education is called "cultural transmission." This is the western traditional stream. Believers in this stream hold that the basic task of educators is "the transmission to the present generation of bodies of information and of rules or values collected in the past; they believe that the educator's job is the direct instruction of such information and rules."[59] To a large extent this understanding dominates most contemporary learning with very few exceptions, generally found in innovative research.

The third understanding of education is progressivist. Progressivism holds that "education should nourish the child's natural interaction with a developing society or environment." Believers of this stream of education "define development as a progression through invariant, ordered sequential stages. The educational goal is the eventual attainment of a higher level or stage of development in adulthood, not merely the healthy functioning of the child at a present level."[60] According to this theory integration of the child within the supportive society is integral. It facilitates personal development without suppressing autonomy. This kind methodology is in agreement with Ubuntu Philosophy.

Kohlberg first suggests that the objective of education be "identified with development, both intellectual and moral." This suggestion is based on his observation of necessarily developmental and progressive nature of education. He therefore posits that "education so conceived supplies the conditions for passing through an order of connected stages."[61] In his view, education should not be internalization of bodies of knowledge but conceptualization of principles. Kohlberg concludes that "a notion of education for development and education for principles is liberal, democratic, and nondoctrinative.

---

[58] Kohlberg,(1981, p. 51).

[59] Kohlberg (1981, p. 52).

[60] Kohlberg (1981, p. 54).

[61] Kohlberg (1981, p. 94).

It relies on open methods of stimulation through a sequence of stages, in a direction of movement that is universal for all children. In this sense, it is neutral."[62] Although cognitive and moral development is a realization of an innate potential, social interaction is a necessary environment for the actualization of the potential. Actually, morality is naturally relational. "Every child believes it is bad to kill because regard for the lives of others or pain at death is a natural empathic response, although it is not necessarily universally consistently maintained."[63]

Social dimension is crucial in Kohlberg's theory of moral and cognitive development essentially, as he states himself, because "developmental theory assumes formalistic criteria of adequacy, the criteria of levels of *differentiation* and *integration*. In the moral domain, these criteria are parallel to formalistic moral philosophy's criteria of *prescriptivity* and *universality*." In other words differentiation and integration are at the core of both cognitive and moral development since they are the ones which make moral and cognitive development a process and they are represented by the society. When combined "the criteria of *prescriptivity* and *universality* represent a formalistic definition of the moral, with each stage representing a successive differentiation of the moral from the nonmoral and more full realization of the moral form."[64]

Obviously, therefore, even though the potential remains within the subject, the process of moral and cognitive development is conditioned by, and contingent on society, social order and personal relationship with both with the society. Kohlberg states,

> Although there are major theoretical differences among sociological role theorists, psycho-analytic theorists, and learning theorists, they all view moral development and other forms of socialization as 'the process by which an individual, born with behavior potentialities of an enormously wide range, is led to develop actual behavior confined within the much narrower range of what is customary and acceptable for him according to standards of his group.'[65]

Kohlberg's stages of cognitive and moral development cannot be understood independent of society. He argues that his study "has indicated the feasibility of looking at individual differences in morality as representing a sequence of stages in conceptualizing the social order and the self's relation to it."[66] Kohlberg's argument indicates that greater social participation and responsibility is indicative of greater moral maturity. In his own words: "we felt in terms of the informal, and statistical analyses, we carried out, that we could not in a general way narrow down our interpretation beyond saying that greater social participation and responsibility in general is related to greater moral development in general."[67]

---

[62] Kohlberg (1981, p. 95).

[63] Kohlberg (1981, p. 15).

[64] Kohlberg (1981, p. 171).

[65] Kohlberg (1981, p. 105). Kohlberg cites Child, I. 1954. Socialization. In *Handbook of social psychology*, ed. G. Lindzey. Reading, Massachusetts: Addison-Wesley.

[66] Kohlberg (1958, p. 358).

[67] Kohlberg (1958, p. 339).

Due to the fact that there is personal potential seeking autonomy and the necessary and inescapable social frame within which the individual is allowed to develop, there is tension between the individual and the society in the process of differentiation and integration. Kohlberg argues that "only a third ideology can resolve the conflict between the society and the individual as the determinant of moral values. The ideology is progressive interactionism, which escapes the trap of either indoctrination or relativism." Progressive interactionism reconciles the process of differentiation and integration as a necessary means to an end. Kohlberg states that "such ideology is philosophically sound because it first rationally attempts to define and justify what should be the ends of education. Moreover it is psychologically sound because it is supported empirically by cognitive developmental research."[68]

In sum Kohlberg's theory of moral development demonstrates that moral development is not independent of cognitive development. That is why Kohlberg's types "reflect, on the whole, both an order of increasing internalization and an order of increased cognitive adequacy or 'rationality' in the moral area." There is correlation and interdependence between the two aspects of development, even though neither can be reduced to the other. "The course of moral development in our data does not seem to be describable in separable cognitive and affective areas. However, this does not imply that the growth of morality is the growth of intelligence—our correlations with intelligence contradict this."[69] Secondly, Kohlberg's theory of moral development attempts to show morality as a dimension of development which "could not be reduced to growing cognitive skill in manipulating value clichés and in anticipating consequences." The theory shows as well that moral development "could not be reduced to learning of 'internalizing' the 'right' values as a readymade set of preferences."[70] The social and the moral aspects of development cannot be separated from each other since "the moral as a fundamental dimension of social development."[71] Essentially, morality is shared by all humans even as it remains transcendental to all. Kohlberg states that "morality introduces a dimension of conformity common to all groups and transcending all. The trends of moral development we have sketched may provide a key to the developmental integration in the individual of the multiple groups to which he belongs."[72]

Kohlberg's theory of education is observably a marriage of the Romanticism of Rousseau, progressivism and some cultural transmission principles of education. However, the theory is much more reliant on Romanticism and progressivism than on cultural transmission. Education, in Kohlberg's view, is a kind of initiation which allows the individual to realize his potential without transgressing societal regulations. In Kohlberg's view, education is both active participation within favorable environment and a definition of societally established boundaries beyond which

---

[68] Kohlberg (1981, p. 4).

[69] Kohlberg (1958, p. 355).

[70] Kohlberg (1958, p. 354).

[71] Kohlberg (1958, p. 337).

[72] Kohlberg (1958, p. 356).

the student cannot go. Therefore "a more complete approach implies full student participation in a school in which justice is a living matter."[73]

The process is simultaneously limited by, and dependent on, the society in which the individual person is located. Apparently, Kohlberg foresees the possibility of an ideally mature person to transcend societal limitation in stage 6 into an ideal (projected) stage seven. Presumably, at that ideal stage an ethically mature person acts both within the framework of societal limitation while taking the whole society to a higher ideal level. In Ubuntu such a person is allowed to act freely because he has proven to be so mature that his action proceeds from his care, not just for the self but also for the society in general.

To a large extent Ubuntu agrees with Kohlberg's theory of cognitive development, which is linked with personal moral development. Ubuntu accepts the fact that there is a potential within people that needs to be stimulated and given the right environment to realize itself. However Ubuntu believes as well in the handing over of the cumulative wisdom that the society has gained over its survival. The most important aspect of moral and cognitive development though, is the interactive one. The potential that a person has for cognitive and moral development will only be realized in the actual practical interaction with the society and environment. It is also within the interactions that cultural wisdom and experience that helps one mature is shared. Noteworthy is the fact that Ubuntu philosophy is praxis based.

3.1.2.3 The Ethical Principle of Justice as Ultimate End of Moral Development

Kohlberg's theory of moral and cognitive development is based on Platonic and Aristotelian philosophies. Kohlberg cites Plato and Aristotle on issues like moral education, moral development, meaning of virtue, and its innate nature. Although Kohlberg formally disagrees with Plato, he nevertheless agrees with the Platonic understanding that teaching of virtues is not a mere instruction. Kohlberg accepts the view that "virtue is ultimately one, and it is always the same ideal" for all cultures. Justice is the ideal form of virtue. "Virtue is knowledge of the good," and since ultimate good is one for all people, virtue in general, and justice in particular is universally accessible by all humans. Good can be taught because "we know it all along dimly," so teaching is "more a calling out than an instruction." The good may not be easily taught under some circumstances because "the same good is known differently at different levels and direct instruction cannot take place across levels." Thus, virtue cannot be taught by impartment. It is truly taught by "asking of the questions and the pointing of the way, not the giving of answers." Thus, moral education is like a labor process whose purpose is "leading of people upward," rather than "putting into the mind of the knowledge that was not there before." Education,

---

[73] Kohlberg (1981, p. 48).

therefore, is autonomous process of growth, a development towards the ultimate good, understanding of which is virtue. In its ideal form though, virtue is justice.[74]

Kohlberg argues that the principle of justice is central, not only in moral development but also in cognitive development. He argues that psychologically "both welfare concerns (role taking, empathy) and justice are present at birth of morality and at every succeeding stage." Only justice, however, "takes on the character of a principle at the highest stage of development." Justice eventually "takes precedence over law and other considerations, including welfare." According to Kohlberg other principles "do not work, either because they do not resolve moral conflicts or because they resolve them in ways that seem intuitively wrong."

Justice is "the only one that 'does justice to' the viable core of lower stages of morality." According to Kohlberg, therefore, justice is the ultimate principle of morality. "The reason that philosophers have doubted the claims of justice as 'the' moral principle is usually that they have looked for a principle broader in scope than the sphere of moral or principled individual choice in the formal sense."[75] He states that "if a formalistic definition of moral principle is unjustified, no one has proposed a better definition. And if an equation of moral principle with justice is injustified, no one has proposed a satisfactory alternative."[76] Kohlberg also observes that "Denial that justice is the central principle of morality thus tends to coincide with a refusal to accept a formal deontological concept of morality but is not backed by an alternative positive definition of morality."[77]

Kohlberg distinguishes rule from principle: "a rule says 'don't do that,' or 'do that,'—it prescribes an action. A principle is some 'rule' which tells us how to make a choice between two more or less legitimate or ruleful alternatives."[78] In other words, rules are prescriptive. They operate within principles. Thus, principles are "neither rules (means) nor values (ends)." They are guides that perceive and integrate "all the morally relevant elements in concrete situations. They reduce all moral obligations to the interests and claims of concrete individuals in concrete situations; they tell us how to resolve claims that compete in a situation, when it is one person's life against another's."[79] Principles therefore transcend concrete situations and the rules applicable in those situations. Kohlberg states that "besides regularity or consistency in use of a reason for choice, a principle implies the universality and ideality of such reason. The basis of choice is one which it would be desirable for all to use."[80]

The principle of justice makes it possible to execute human rights. Kohlberg defines human rights as "a claim for some positive action by another. It is a legitimate

---

[74] Kohlberg (1981, p. 30).
[75] Kohlberg (1981, pp. 175–176).
[76] Kohlberg (1981, p. 177).
[77] Kohlberg (1981, p. 176).
[78] Kohlberg (1958, p. 287).
[79] Kohlberg (1981, p. 175).
[80] Kohlberg (1958, p. 288).

expectation as to the actions of other persons or of the social system."[81] Claims are relational in nature as they involve two parties. The language of rights, therefore, calls for the principles of justice. According to Kohlberg "by definition, principles of justice are principles for deciding between competing claims of individuals, for 'giving each person his due.' When principles, including considerations of human welfare, are reduced to guides for considering such claims, they become expressions of the single principle of justice."[82] Consequently, "The most basic principle of justice is equality: treat every man's claim equally, regardless of the man."[83] Without acknowledging human equality most moral principles remain baseless.

Basing his thinking and logic on Platonic and Aristotelian philosophy, Kohlberg makes a general conclusion that "the man who understands justice is more likely to practice it…youths who understand justice act more justly, and the man who understands justice helps create moral climate which goes far beyond his immediate and personal acts. The universal society is the beneficiary."[84] However, moral principles are general by nature and their "generality cannot be coercive…one can never coerce others to think or decide in any given way…accordingly, a principle of choice must appeal to 'reason,' for its acceptability. It must seem to command assent intrinsically."[85] Moreover, Kohlberg asserts, "moral judgments, unlike judgments of prudence or esthetics, tend to be universal, inclusive, consistent, and grounded on objective, impersonal, or ideal grounds."[86]

Kohlberg refers to Socrates and Martin Luther King as teachers of justice who put their teaching in practice regardless of the cost. They died for justice.[87] Martin Luther King and Socrates knew the good and pursued the ultimate good regardless of the impediments. They could not avoid doing the good. According to Kohlberg, such people were really mature in their understanding of justice. They transcended and surpassed their respective communities in their pursuit of justice. "King makes it clear that moral disobedience of the law must spring from the same root as moral obedience of the law, of respect for justice" because there cannot be contradiction with regards to *the good*, justice or rights. "We respect the law because it is based on rights both in the sense that the law is designed to protect the rights of all and because the law is made by the principle of equal political rights. If civil disobedience is to be stage 6, it must recognize the contractual respect for law of stage 5, even to accepting imprisonment."[88]

Once again, Kohlberg, like Plato, uses generality of "*the good*" and justice to relate knowledge and practice. According to this understanding, one who knows the good will seek it. To the degree one knows *the good*—one will pursue it. Justice

---

[81] Kohlberg (1958, p. 252).

[82] Kohlberg (1981, p. 175).

[83] Kohlberg (1971, p. 51).

[84] Kohlberg (1968a, p. 30).

[85] Kohlberg (1958, p. 289).

[86] Kohlberg (1981, p. 170).

[87] Kohlberg (1981, p. 401).

[88] Kohlberg (1981, p. 43).

follows from the universal nature of *good* and *right*. Justice therefore "must appeal for acceptance on the grounds that it is appropriate for 'a reasonable being' to adhere to."[89] There is a difference between understanding of a virtue or principle and its practice. Practice of a principle is based on, and affected by, many other factors than merely understanding it. Such factors like emotion, uniqueness due to different personalities, and concrete situation impact practice of justice.

In sum, the end or objective of all cognitive and moral development in Kohlberg's theory as it is in Ubuntu seems to be societal, biospheric and of the cosmic good. Kohlberg indicated that there is a tendency for someone who is really mature, one who really lives a just life to transcend the principle of justice because "what empowers a person to live a life of justice, and to face death for it, is itself something 'beyond justice,' something I metaphorically call 'stage 7.'" Thus moral and cognitive development is a continuum. Moral maturity is not really fully attained. It tends to become progressively all-embracing and universal as it enters into stage seven. Kohlberg states that people at stage 7 "affirm life from a 'cosmic perspective'; feel some mystic union with God, Life, or Nature; and accept the finitude of the self's own life, while finding its meaning in a moral life, a life in which a sense of love for, and union with, Life or God is expressed in a love for fellow human beings."[90] Stage 7 of Kohlberg's theory of cognitive and moral development has been demonstrated as the ideal of Ubuntu culture. From the perspective of Ubuntu, Kohlberg's stage 7 is attained through caring, not justice. Care in Ubuntu is a much higher ethical value than justice.

3.2 Human Relationships

While Kohlberg may be justified to contend that the majority of moral problems arise from competing rights and that their resolution depends on conception of human rights, resolution of moral problems cannot ignore the parties' respective location in the six stages of moral development because people perceive justice differently at different stages of moral development. If perception of justice is compromised by the six moral development stages and if very few people get to the sixth stage, moral objectivity and universality of values and principles are at least compromised and at most relativized by the stages. This assertion calls for further research. Carol Gilligan disagreed with Kohlberg's criterion of moral development based solely on perception of justice.

Later Kohlberg acknowledged that morality may not be based solely on justice and human rights. He acknowledged that Gilligan captured "a second sense of the word *moral*" which focuses on care and responsibility.[91] Thus the second major component of Ethics of Care concerns human relationships. Individual and universal rights need to be interpreted in light of ethical responsibility having meaning

---

[89] Kohlberg (1958, p. 289).
[90] Kohlberg (1981, p. 401).
[91] Kohlberg Lawrence. The current formulation of the theory, 87.

within human relationships. This component has two related concepts. First, self in relationship according to Carol Gilligan's three stages of moral development; second, ethics and narrative based on Nel Nodding's perspective on uniqueness of each moral problem and the need for personal contact. Human relationship is central to Ubuntu, which is why Ubuntu has a lot in common with ethics of care.

3.2.1 Self in Relationship

The first concept of human relationships is based on the self in relationship. Reacting to Kohlberg's theory of moral development, Gilligan asserts that moral development proceeds from egocentrism to an-other-oriented stance and culminates in a final stage in which the self in relationship with another comes into balance.[92] Gilligan argues that Kohlberg's theory of moral development was not impartial, that it excluded women and an important aspect of moral development. Kohlberg's theory excludes the role of human relationship and responsibility. Gilligan's theory of moral development has its foundation in Kohlberg's theory of moral development but then expands upon it. According to Gilligan, there are three stages of moral development: pre-conventional, conventional and post-conventional.

The goal of pre-conventional stage is individual survival. The goal of the conventional stage is responsibility to others (self-sacrifice is goodness). The goal of the post-conventional stage is truth (That is a person too!). The major difference between Kohlberg and Gilligan's theories of development is that in Gilligan's theory the transition is fueled by a change in the sense of the self, while in Kohlberg the transition is fueled by changes in cognitive ability.[93]

Ubuntu ethics identifies with Gilligan's theory of moral development in the a number of ways, the first of which being that both in Ubuntu and in Gilligan's theory of moral development, relationship is essential and that development is facilitated not by competition but by self-perception in relation to the society ("the other"-using Ubuntu language). In both Gilligan's theories and Ubuntu, responsibility towards others is crucial. Maturity involves self-sacrifice for others and that recognition of the equality of the other to the self is essential.

3.2.1.1 Buber's *I and Thou* as an Inspiration to Care Ethics

There is a lot in common between Buber's existentialism and Ubuntu ethics. The two philosophies are about interdependence between *I and Thou*. The "I" stands always in the presence of the "Thou." A human being realizes that his life is interlocked and contingent on other human lives and other reality around him a posterio-

---

[92] Gilligan Carol. 1982. In a different voice: Women's conceptions of self and morality. *Review of General Psychology* 6 (2): 139–145.

[93] Gilligan (1982).

ri. That means the *I—Thou* relationship is a priori, a condition for human existence. Exploring human language, Buber refers to human existence's contingency to other existents. "If *Thou* is said, the *I* of the combination *I-Thou* is said along with it. If *it* is said, the *I* of the combination *I-It* is said along with it." In other words, human language reveals nature of reality; that is, reality's ontological unity. Buber elaborates this fact when he writes, "There is no *I* taken in itself, but only the *I* of the primary word *I-Thou* and the *I* of the primary word *I-It*…For where there is a thing there is another thing. Every *it* is bounded by others; *it* exists only through being bounded by others."[94] Buber's view of reality reflects Ubuntu philosophy

Buber's use of the *I-Though* language underlines the importance of relationships between humans and between human beings and the rest of reality. He states, "The primary word *I-Thou* establishes the world of relation."[95] It is through such relationship that the self becomes self. "Through the *Thou* a man becomes *I*. That which confronts him comes and disappears, relational events condense, then are scattered, and in the change consciousness of the unchanging partner, of the *I*, grows clear, and each time stronger." Any human being, however, remains constantly, as Buber puts it, "caught in the web of the relation with the *Thou*, as the increasingly distinguishable feature of that which reaches out to and yet is not *Thou*." It is the *I-Thou* relationship, which enables even self-examination or introspection. Buber states that the *I-Thou* relationship facilitates consciousness and development as it "continually breaks through with more power, till a time comes when it bursts its bonds, and the *I* confronts itself for a moment, separated as though it were a *Thou*; as quickly to take possession of itself and from then on to enter into relations in consciousness of itself."[96]

There is no doubt, therefore, that individuality develops from relationships. Relationships, however, indicate human neediness of preexistent others, with whom to relate both for his own psychological, moral and social development and for their development. Such neediness for others is basic and, as it were, a prerequisite of personal development. In Buber's words, "The person becomes conscious of himself as sharing in being, as co-existing, and thus as being. Individuality becomes conscious of itself as being such-and-such and nothing else. The person says, 'I am,' the individual says, 'I am such-and-such.'" Consequently, personal consciousness is self-definition or distinction from the other (*Thou*). In other words, as Buber puts it, "'Know thyself,' means for the person 'know thyself to have being,' for the individual it means 'know thy particular kind of being.' Individuality in differentiating itself from others is rendered remote from true being."[97]

Since reality is essentially a unity, or so to say, an organism. As much as individuation and individuality is important for any relationship, it should be controlled by reality for it depends on it. Buber writes, "The more a man, humanity, is mastered by individuality, the deeper does the *I* sink into unreality. In such times the person

---

[94] Buber (1958, pp. 3–4).

[95] Buber (1958, p. 6).

[96] Buber (1958, pp. 28–29).

[97] Buber (1958, pp. 63–64).

in man and in humanity leads a hidden subterranean and as it were cancelled existence—till it is recalled."[98] If this is the case, there is a need for balance between individuality and its relationship with reality as a whole. This balance is important because "Every real relation in the world rests on individuation, this is its joy—for only in this way is mutual knowledge of different beings won—and its limitation—for in this way perfect knowledge and being known are foregone."

Cognitive awareness does not only reveal interdependence of reality but also human limited control of reality and his exposure to it. Buber points to this when he writes, that "in the perfect relation my *Thou* comprehends but is not myself, my limited knowledge opens out into a state in which I am boundlessly known."[99] The ontological and cognitive interaction between an individual and the rest of reality is transformative, which means, it is in a state of flux or constant change. "Every real relation in the world is consummated in the interchange of actual and potential being; every isolated *Thou* is bound to enter the Chrysalis state of the *It* in order to take wings anew."

The change in individuals, which is inevitable, is based on relation and it is ongoing. However, "in pure relation potential being is simply actual being as it draws breath, and in it the *Thou* remains present."[100] Thus, no individual human can exist as human independent of other humans and the cosmos, just as no individual can resist being simultaneously actual and potential, due to the fact of human inevitable interaction with reality and its consequential change.

Buber's existentialism verbalize in a very realistic way Ubuntu philosophy represented in maxims: *I am because you are; I am who I am because you are who you are; I am because we are; a human being is a human being because of other human beings; a human being is a human being because of the otherness of other human beings.* All that the maxims explain is the fact that reality is an organism, a unity in plurality. The plurality and diversity within the essentially unified reality enables individualization and its realization. Arguing for individual rights that do not recognize other individuals' equal rights is *reductio ad absurdum*.

3.2.1.2 Mcquarrie's *Existentialism* as Care Ethics' Worldview

Exploration of Ubuntu reveals that it is essentially existentialism, not substantially different from the one found in the philosophy of Mcquarrie. Ethics of care is equally based on existentialism. Existentialism has either directly or indirectly influenced the discourse and the perspective of ethics of care. Like Buber, Macquarrie observes and asserts, "The existent lives are in constant interaction with other existents." Macquarrie categorically states, "Existence is 'being-with-others' or 'being-with-another.'" Macquarrie's statement indicates existence is contingent to otherness. Ethical principles such as autonomy, beneficence, non-maleficence and justice are

---

[98] Buber (1958, p. 65).

[99] Buber (1958, pp. 99–100).

[100] Buber (1958, p. 100).

based on the assumption of otherness without which they are rendered meaningless and irrelevant. This underlines what Macquarrie says, that is, "existence is fundamentally communal in character, and without the others I cannot exist."[101]

In compliance with Ubuntu worldview, Macquarrie posits that "society is not formed by the banding together of individuals" as it has traditionally been assumed, rather, "Individuals emerge from a society that is prior to them." Macquarrie explains that this assertion has been argued and accepted by most modern anthropologists and sociologists.[102] By his observation and argument on the precedence of society over individuals, Macquarrie positions individuality in realistic balance, exaggerating or undermining which is an ontological and epistemological mistake. He states that "Individualism and collectivism are, at bottom, different forms of the same error. We can avoid them only if we begin with the concreteness of existence as 'being-with-others.'"[103]

In Macquarrie's view, the world and otherness have to be taken seriously because "this world is an a priori condition of all my practical concerns…so one may also claim that the others are a priori—they are conditions of existence rather than 'extras' that are added on to existence." An individual finds himself in a world of many other individual humans and non-humans mutually interrelating in a way that makes individuality possible and realizable.[104] However, only human beings are capable of personal existence and relationship. Macquarrie observes that "no animal, no crystal, no manufactured thing says 'I.'"

The ability to use the personal pronoun *I* defines, not only uniqueness and specialness of a human being but also his different interaction and relationship with the rest of reality. Macquarrie argues that "the uniqueness of the human existence lies in the felt 'mineness' of that existence which knows itself as 'I,' almost a microcosm." A human person is capable of self-reflection. He is aware of his own existence. Macquarrie states that a human being is a center "different from every other, at once lonely and cut off, yet also in a sense embracing the world and embraced by it."[105]

Due to humans' unique capability of relating with their environment, humans are not only capable of consciously effecting change in their environments; they are capable of evaluating the change that they effect. With regards to their relationship with other humans, Macquarrie posits that a human's being with others ought to be authentic: "Authentic being-with-others is precisely that mode of relation to the other that promotes existence in the full sense; that is to say, it lets the human stand out as human, in freedom and responsibility." Therefore, being with others authentically means being responsible in recognizing and respecting the other in his uniqueness. It also means recognizing the inviolable rights of the other. "On the other hand, inauthentic being-with-others suppresses the genuinely human and personal. Whatever kind of relation to the others depersonalizes and dehumanizes

---

[101] Macquarrie (1972, p. 102).

[102] Macquarrie (1972, p. 103).

[103] Macquarrie (1972, p. 104).

[104] Macquarrie (1972, p. 104).

[105] Macquarrie (1972, p. 74).

is an inauthentic one…True community allows for true diversity."[106] Macquarrie implies that failure to recognize other humans' personhood and its rights and obligations is falsifying reality, which in turn makes our relationship with such humans both self and other-deceptive, in other words, unethical. Unethical relationship or treatment of the other is thus inauthentic deceptive and falsifying.

Representing an existentialist perspective, Macquarrie objects to the Cartesian perspective which, in his view, exaggerates the importance of thought, especially because thinking is secondary to existence. There can be existence without rationality. Macquarrie argues, "I am not primarily a thinking subject. I am first of all an existent: existence is something much broader than thinking, and prior to it."[107] While existence is of utmost importance, its dependence on its environment is crucial. Like Buber, Macquarrie recognizes dependability of human existence on its immediate environment and the cosmos.

He simply but categorically states, "Existence is being-in-the-world, and there is no existence without environment."[108] With regards to human relationship Macquarrie recommends genuineness. Ideal human relationship is essentially mutually affirmative, mutually respectful, mutually equal and mutually reciprocal. "A genuine relation to another person cannot be one-sided, dominating, or possessive; it must consist in openness and willingness to listen and receive as well as to speak and to give."[109] Thus, as humans find themselves already related to other humans, their environment and the cosmos, they have an ethical duty to genuine human relationship, which by nature is incapable of reducing the other into anything less than human.

Clearly Macquarrie's conception of reality is almost a replication of the Ubuntu philosophy. The only difference is the fact that Macquarrie's existentialism was put to writing while Ubuntu which had been in existence prior to human ability to writing was passed on by word of mouth and practice.

3.2.1.3 Gilligan's Theory of Women Cognitive and Moral Development Based on Care

According to Kohlberg, individual human beings have basic inalienable rights, which all other humans should respect. The role of morality, therefore, is to define boundaries and impose restrictions in order to protect those rights. Thus morality is concerned with justice in the allocation or recognition of individual rights and in the protection of the defined boundaries of individual rights. This perspective dominates modern Western ethics. However, Gilligan comes up with an important dimension of cognitive and moral development that Kohlberg overlooked, that is, care and human mutual responsibility for one another.

---

[106] Macquarrie (1972, p. 121).

[107] Macquarrie (1972, p. 125).

[108] Macquarrie (1972, p. 93).

[109] Macquarrie (1972, p. 109).

According to Gilligan, a female approach to morality is based on relationship and responsibility. The main assumption in this theory is that individuals have responsibility toward other individuals. Morality should be concerned with individual responsibility and care for other individuals. Gilligan assigned this perspective to the female gender. This perspective is almost opposite to that of men: men desire to limit interference (desire for separation and fear of commitment) while women desire for meaningful harmonious connections and commitments.[110]

In the first stage of Gilligan's theory of moral development, children are preoccupied with individual survival. The importance of the need to survive renders them basically selfish. However, as they develop, children learn to pay attention to what happens to others and eventually learn to empathize with them. Empathy challenges them to start equating their needs and their very selves with others. They gradually start moving away from their selfishness as they develop greater concern for others. This concern for others at stage two (conventional morality stage) is based on recognition of basic human equality.

Empathizing with others tend to be exaggerated before, in reaction, children start realizing that ignoring their own needs for those of others is as equally wrong as ignoring other people's needs. This realization brings them to the final stage (stage three) in which responsibility and care for both self and others is perceived as the moral ideal. Such responsibility is indicative of, and requiring some sacrifice. "The woman at this stage validates her claim to social membership through the adoption of societal values. Consensual judgment about goodness becomes the overriding concern as survival is now seen to depend on acceptance by others." Resolution of moral problems should seek not only care for others but inclusion and minimization of harm.[111] At the final stage there is need for balance between self and others. Selflessness is not ideal as it hurts both the self and her relationships. In case of conflict involving power and care women in the third stage of development would give up power for care.[112]

Gilligan argues that, owing to women's tendency to avoid harm to anyone, which in her view, is a fundamental concern for women, women tend to avoid judging. However, women's reluctance to judge is not moral relativism. She argues that women recognize and take into consideration practical world situations and uniqueness of individual experience which is not easily always reducible to a simple theory or moral principle. Real human situation is unique and complicated. Women tend to consider that fact better than men do.[113] One of the major claims that Gilligan makes concerns relationships. She argues that women differ from men in the way they experience, and deal with relationships, especially dependency. According to Gilligan men develop differently from women. Actually, Gilligan portrays the two trends of development, that of boys and that of girls as opposites:

---

[110] Gilligan, Carol. 1993. *In a different voice: Psychological theory and women's development*, 38. Cambridge: Harvard University Press.

[111] Gilligan (1982, pp. 79–80).

[112] Gilligan (1982, p. 95).

[113] Gilligan (1982, pp. 100–105).

> For boys and men, separation and individuation are critically tied to gender identity since separation from the mother is essential for the development of masculinity. For girls and women, issues of femininity or feminine identity do not depend on the achievement of separation from the mother or on the progress of individuation. Since masculinity is defined through separation while femininity is defined through attachment, male gender identity is threatened by intimacy while female gender identity is threatened by separation. Thus males tend to have difficulty with relationships, while females tend to have problems with individuation.[114]

Apparently, at least from Gilligan's view point, this argument explains the difference between male and female perspective on morality. Women go on defining themselves in terms of, and in the context of human relationship and ability to care, while men tend to perceive relationship and care as weakness which rivals autonomy and independence. Gilligan asserts that "when the focus on individuation and individual achievement extends into adulthood and maturity is equated with personal autonomy, concern with relationships appears as weakness of women rather than as a human strength."[115] In the case of women, however, the case is different. Ability to relate, engage and care is not a weakness but strength. Care and mutual responsibility neither compromise nor threaten ethical autonomy. According to Gilligan's study for women "obligation and sacrifice override the ideal of equality," thus creating conflict between care and justice.[116]

Like Buber and Macquarrie, Gilligan asserts that the self needs the other in a fundamental, existential and ontological way. Gilligan states that "the truth of relationship, however, return in the rediscovery of connection, in the realization that self and other are interdependent and that life, however valuable in itself, can only be sustained by care in relationships."[117] The main difference between Buber's and Macquarries assertion against Gilligan's assertion is that Buber and Macquarrie make universal and categorical statement while Gilligan's statement is biased towards women care.

According to Gilligan, "in all the women's descriptions, identity is defined in a context of relationship and judged by a standard of responsibility and care." This general statement about women, however, seems to exclude men. One wonders whether men's identity can be defined independent of relationship. Gilligan goes on to assert that women perceive morality "as arising from the experience of connection and conceived as a problem of inclusion rather than one of balancing claims."[118] Gilligan's conclusions with regards to her distinction of women's against men's perception of morality seem to be sweeping and overgeneralizing. This problem shows that she is as biased as Kohlberg, though for women and against men.

Gilligan's theory of moral development is in agreement with the philosophy of Ubuntu except for its overly association of care and relationship with women, and

---

[114] Gilligan (1982, p. 8).
[115] Gilligan (1982, p. 17) citing Miller Jean Baker. 1976. *Toward a new psychology of women.* Boston: Beacon Press.
[116] Gilligan (1982, p. 64).
[117] Gilligan (1982, p. 127).
[118] Gilligan (1982, p. 160).

autonomy with men. As a philosophy Ubuntu is gender neutral and care is a moral ideal for both men and women. Relationship belongs to the very kernel of being human. Ubuntu philosophy would not condone Gilligan's perspective, as it is either culture-conditioned or too limited to developmental psychology genre. The psychological analysis that Gilligan provides in reaction to Kohlberg's theory of cognitive and moral development contradicts, not only Kohlberg's theory of moral development, it also contradicts Buber's existentialism and Ubuntu.

3.2.2   Ethics and Narrative

According to Ubuntu philosophy as it is with most care ethicists, morality is concerned with the activity of care. Moral problems arise from conflicting responsibilities; thus their resolution should be practical, contextual and narrative.[119] Noddings argues that "Since so much depends on the subjective experience of those involved in ethical encounters, conditions are rarely 'sufficiently similar' for me to declare that you must do what I must do."[120] This approach towards care would imply that care ethics does not stipulate any substantive norms, but rather consists of an attitude of attending to the other's wants and needs.

Even though Noddings argues that care represents a universal morality, she claims it occurs only in intimate relations where it is highly variable and subject to the practical judgments of the care-giver. Consequently, Noddings concedes that her meaning of care entails a particular situational morality. We may "care about" strangers in the sense of maintaining "an internal state of readiness to try to care for whoever crosses our path, but she distinguishes this perspective from 'caring-for' to which we refer when we use the word 'caring'"[121] Caring per se requires personal contact and varies according to individuals and situations. Indeed, because of the particularity of care, Noddings is wary of passing judgment on the caring activities of others. What is good for one individual in one situation may not be good for another in another situation.

According to Ubuntu, care can, and should always be generalized to include everybody and the entire human environment. Noddings particularization of care reduces care to personal intimate relationships, which cannot be universalized into a philosophy. As such, care is not an ethic. The Ubuntu perspective of care is just the opposite of Noddings' perspective. In other words, according to Ubuntu cognitive and morals, maturity is directly associated with one's ability to transcend natural tendency to care only for one's intimates into caring for all and about common good.

---

[119] Ward et al. (1988, p. 21–48).

[120] Noddings (1986, p. 5).

[121] Noddings (1986, p. 18).

3.2.2.1 Derrida on the Injustice of Law

According to Derrida, There is an injustice in the process of law. Such injustice is based on overriding of individuality and its multi-dimensional contexts, all individuals and their contexts being unique, thus different from any other. The law overgeneralizes by treating everybody more or less equal with everybody else. Real morality recognizes and deals with "singularity, individuals, irreplaceable groups and lives, the other or myself as other, in a unique situation."[122] In his view, justice is generally uncaring. Strictly speaking, caring justice is impossible even between intimates.

Caring justice would be only possible if an individual would be able to assume another person's existence fully so that he could address the needs of the cared-for person authentically, truly and really. Derrida states, "To address oneself to the other in the language of the other is, it seems, the condition of all possible justice." Unfortunately, it is impossible since "I cannot speak the language of the other except to the extent that I appropriate and assimilate it according to the law of an implicit third."[123]

Derrida's concern is really finding fairness in justice as practiced in civil law. In trying to discern justice in the practice of the law, Derrida finds out that the law may be unjust and irreparably so. Like Noddings, Derrida holds that moral ideal consists of paying "infinite attention to the needs and perspectives of others." Since giving infinite attention to another person is attainable, there is a sharp distinction between "caring justice and the exercise of justice as law or right, legitimacy or legality, stabilizable and statutory, calculable a system of regulated and coded prescriptions."[124] One may rightly state that there is inevitable irreparable injustice in the practice of civil justice if one thinks of 'caring justice' using the language of Derrida. However, warns Derrida, infinity of justice "cannot and should not serve as an alibi for staying out of juridico-political battles, within an institution or a state or between one institution or state and another."[125] Even if there is room for improvement within the law, the ideal of the law cannot be achieved. Real caring justice in its ideal form remains an inspirational ideal.

With human inability to achieve perfect justice, the ideal of justice should remain an ideal while praxis should strive to realize it as much as that is possible at any particular time in any specific issue. The best way to do it according to Derrida is to "hear, read, interpret it, to try to understand where it comes from, what it wants of us, knowing that it does so through singular idioms." Derrida then recommends "never to yield on this point, constantly to maintain an interrogation of the origin, grounds and limits of our conceptual, theoretical or normative apparatus surrounding justice."[126]

---

[122] Derrida (1990, p. 949).
[123] Derrida (1990, p. 494).
[124] Derrida (1990, p. 959).
[125] Derrida (1990, p. 971).
[126] Derrida (1990, p. 955).

Through philosophical ethics and politics, Derrida addresses the ideal of Ubuntu which is represented in the ideal of care: identification with the cared-for. In Ubuntu, however, the ideal does not remain abstract but always made present so that it constantly inspires and challenges each individual. In Ubuntu the ideal is represented by the maxims, *I am who I am because you are who you are*; and *human beings are human because of the otherness of other human beings*. Both statements, when explained not only inspire caring justice, they challenge care-givers to identify with recipients of care while, at the same time, respecting and not compromising their unique identity and singularity. One important difference between Ubuntu and Derrida's perspective on law is the fact that Ubuntu recognize the importance of Universal law. According to Ubuntu law is an instruction and guidance on how to care. Personal uniqueness does not nullify objectively proven universal principles and laws. Ubuntu discourages dictatorship of moral relativism.

3.2.2.2 Noddings' Argument for Contextual, Particular and Symbiotic Nature of Care

According to Noddings care is fundamental and universal because, in her view, it is an "attitude which expresses our earliest memories of being cared for and our growing store of memories of both caring and being cared for." Since nobody could ever survive without being cared for by others, care is "Universally Accessible." Thus, care as such cannot be rejected.[127] In her view, caring involves "stepping out of one's own personal frame of reference into the other's." Thus care is empathic and sacrificial because, as Noddings puts it, "when we care, we consider the other's point of view, his objectives, needs, and what he expects of us. Our reasons for acting, then, have to do both with the other's wants and desires and with the objective elements of his problematic situation."[128]

Noddings interpretation of care, however, is relativized by proximity, kinship and geographical distance. Although people have, as they should, caring attitude, their care for strangers is limited. For strangers she uses the phrase "care about" while for those close to, the caring subject she uses the phrase "caring for."[129] From Nodding's perspective, real care cannot be universalized. She rejects "the notion of universal caring—that is, caring for everyone—on the grounds that it is impossible to actualize and leads us to substitute abstract problem solving and mere talk for genuine caring."[130]

Noddings advocates for fundamental moral relationship between human individuals, their immediate environment and the cosmos in a plausible and convincing style which is very similar to that found in Ubuntu. She views human beings' immediate environment as an extension of their bodies. Humans interact with their

---

[127] Noddings, Nel. 1984. *Caring: A feminine approach to ethics and moral education*, 5. Berkeley: University of California Press.

[128] Noddings (1986, p. 24).

[129] Noddings (1986, p. 18).

[130] Noddings (1986, p. 18).

space in a way that there is undeniable mutual influence which transforms both the environment and the specific human person in it. She mentions "houses, rooms and corners as extensions of our bodies; gardens as immediate spaces between home and wilderness or city." She concludes, as is the case with Ubuntu that "place becomes part of the developing self and, in the extreme, the self may even become inextricable from its physical place. Place does not *determine* the self, but it influences and shapes it."[131] Thus Noddings contends that human beings cannot be dissociated from the environment that they interact with. This argument is far reaching since it points to the importance of psychological, geographical, chronological, emotional and sociological environment in which human action happens. Since human being cannot be really extricated from such contexts, such contexts must be taken seriously in morality because humans have a symbiotic relationship with their environment.

There is a communication between human subjects and their environment. Humans learn to respond to their environment. The habit to respond is acquired at home. It is "directed at animals, plants and objects encountered there." The capacity to respond that humans develop at home "develops a basic moral need—the need to care is revealed and, with it, there is a move beyond duty to something deeper" which "induces the great joy of reciprocity."[132] Noddings points out that "the most important entities in early life are other selves. Even in intellectual life, it is not so much objects and buildings that shape us as it is other intellects."[133] Human beings learn how to respond to "the needs of the cared-for," such needs "are captured by the response 'I am here.'" However, some needs, though legitimate, remain unmet due to limitations in resources or conflict.[134]

Due to the uniqueness of each person and context Noddings argues that there cannot be identical or similar moral situations. Ethical encounters being unique, general principles are rendered irrelevant. In Noddings words, "Since so much depends on the subjective experience of those involved in ethical encounters, conditions are rarely 'sufficiently similar' for me to declare that you must do what I must do."[135] Even though Noddings argues that her position does not make morality relative, she actually does argue for moral relativity. She makes ethical principles relative to individuals, their experience and their context. This principleless stance towards morality is reflected in statements like: the goal "lies in trying to discern the kinds of things I must think about" in caring for others.[136]

While one cannot rule out the role of context and individual perspective in ethics, one cannot rule out ethical principles. Doing that would render ethics and morality devoid of meaning and substance, which situation would lead to moral anarchism. Actually, individual positions in any moral situation are, and should be based on ontologically, socially or culturally established and accepted principles. Although

---

[131] Noddings (2002, p. 174).

[132] Noddings (2002, pp. 174–175).

[133] Noddings (2002, p. 175).

[134] Noddings (2002, p. 247).

[135] Noddings (1986, p. 5).

[136] Noddings (1986, pp. 13–14).

Ubuntu is about genuine and authentic care, Ubuntu resists lawlessness and relativism. Morality cannot be reduced to subjective judgments of each person. However, Ubuntu justice is always communitarian and reconciliatory. The community would spend an ample time for conflict resolution, which would generally involve a lot of active listening, and negotiation facilitated and supported by informal community setting until there is a sense of fairness and peace. Pursuit of justice is, at the same time, a therapy. Thus justice is about healing and reconciliation. Ubuntu justice is care.

3.2.2.3   Tronto's Meaning of Care and Equality of People as an End Rather than a Means

The philosophy of Ubuntu partially agrees with Tronto that caring involves self-investment and devotion mentally, emotionally, and physically. Care is oriented to praxis and difference making, without which it is devoid of meaning. Tronto states, "Care implies reaching out to something other than the self…care implicitly suggests that it will lead to some type of action."[137]Care can be understood in at least four different interrelated phases: "caring about, taking care of, care-giving, and care-receiving."[138] Caring about is at its basis paying attention to a needy situation, recognizing that the morally needy situation needs to be resolved. Caring about must involve "noting the existence of a need and making an assessment that this need should be met."[139] Taking care of implies taking initiative. It "involves assuming some responsibility for the identified need and determining how to respond to it… taking care of involves notions of agency and responsibility in the caring process."[140] Care giving is the sacrificial and the practical phase of care. It "involves the direct meeting of needs for care. It involves physical work, and almost always requires that care-givers come in contact with the objects of care."[141] Care receiving concerns the recipient of care. "Care-receiving…recognizes that the object of care will respond to the care it receives." This aspect of care helps care-giver "know that caring needs have actually been met."[142] This important phase of care demands some kind of reciprocity.

Tronto assumes that care is natural to people and nature. Tronto shares in the Ubuntu worldview about care, when for instance she suggests that "caring be viewed as a species activity that includes everything that we do to maintain, continue, and repair our 'world' so that we can live in it as well as possible. That world includes our bodies, ourselves, and our environment, all of which we seek to interweave in

---

[137] Tronto (1993, pp. 102–103).

[138] Tronto (1993, p. 107).

[139] Tronto (1993, p. 107).

[140] Tronto (1993, p. 107).

[141] Tronto (1993, p. 107).

[142] Tronto (1993, pp. 107–108).

a complex, life-sustaining web."[143] In other words, for its own survival nature cares for itself; humans need to care for themselves and for the biosphere and the cosmos both for humans beings' own survival and that of the biosphere.

Like the Ubuntu philosophy, Tronto equates morality and moral goodness with care. A moral person is one who "strives to meet the demands of caring that present themselves in his or her life." This assertion would mean that a person who does not care is at best not moral, if not immoral. Tronto goes on to apply the same standard to states and societies. She states, "For a society to be judged as a morally admirable society, it must, among other things, adequately provide for care of its members and its territory."[144] The standard used by Tronto for a moral person and moral society is recognized and applied by Ubuntu ethics. Notably, Tronto's perspective on morality is shared by Ubuntu perspective in a substantial way as explored in Chapter One and Two of the present work.

Tronto falsifies the assumption of equality of people. She holds it as an ideal to be achieved rather than a status quo. She states, "Rather than assuming the fiction that all citizens are equal, a care perspective would have us recognize the achievement of equality as a political goal."[145] The main problem with Tronto's objection lies in the use of the word "equality." Human equality is an ontological and ethical fact that should be recognized as is. It means that all human beings share the same essence of humanity and that at that level they are equal. From the recognition of the equality of essence, human rights proceed. It is on this recognition that ethical principles and theories are founded. However, it is quite true that the ideal of treating every human being as an equal, with equal basic rights with any other human being has never been realized. It remains always as a goal that transcends actual practical life. To that extent, therefore, Tronto is right. She thus concludes in a way that resonates with Ubuntu that, "It is a fact of great moral significance that, in our society, some must work so that others can achieve their autonomy and independence."[146]

3.3   Reciprocity of Care

One of the major components of Ethics of Care concerns reciprocity of care. To integrate the debate on individual/universal rights and relationships, there needs to be reciprocity of care that clarifies the meaning of ethical responsibility. This component has two concepts. First, ethics and context, since according to Virginia Held and Joan Tronto each unique moral situation is located in a unique context; second, the problem of universalization of care based on the views posited by Virginia Held and Nel Noddings.

---

[143] Tronto (1993, p. 103).

[144] Tronto (1993, p. 126).

[145] Tronto (1993, p. 164).

[146] Tronto (1993, p. 165).

3.3.1 Ethics and Context

The first concept of reciprocity of care concerns ethics and context. Ethics of care rejects the dominant moral theories as abstract and ineffective in resolving concrete contextualized moral problems. Each moral problem is different and unique to specific circumstances.[147] Two of the distinguishing elements of ethics of care are emphasis on the concrete and emphasis on the particular. Care ethics takes the concrete needs of particular individuals in specific circumstances as the starting point for what must be done.[148] Ethics of care seriously considers persons as relational rather than separate independent entities.

Consequently, ethics of care "values emotion rather than rejects it."[149] Ethics of care respects and considers the claims and the situation of "particular others with whom we share actual relationships."[150] The role of context and human relationships can hardly be exaggerated in morality. In some sense situational morality is inevitable, owing to the fact of the uniqueness of moral contexts. However, the traditional general principles are crucial in enlightening each unique moral context. Due to its holistic approach to human personhood, moral attitude of mind and importance of human relationships. Ubuntu, like ethics of care, considers human emotion, relationship, mental attitude and intention very seriously in determining morality of human action. However, Ubuntu does not reject general and objective moral principles. It integrates such principles with concrete particularized and contextualized ethical situations. The general principles act as frames and matrices that regulate ethical discernment.

3.3.1.1 Noddings and the Role of History and Context in Moral Development

Ubuntu is in agreement with Noddings argues that ethics is about care. The genesis of ethical caring is, in Noddings' view, psychological and natural to human beings. She describes "ethical caring—the relation in which we meet the other morally," as "that relation in which we respond as one-caring out of love or natural inclination." She further elaborates the relationship of 'natural caring' psychologically as "a human condition that we, consciously or unconsciously, perceive as 'good.'" Noddings contend that human beings long for the condition she describes as 'natural caring.' It is to be in a relation of 'natural caring' that "provides the motivation for us to be moral. We want to be moral in order to remain in the caring relation and to enhance the ideal of ourselves as one caring."[151] Thus Noddings' view of care is that it is a natural ideal

---

[147] Held (2006, p. 11).

[148] Tronto (1993, pp. 102–105); Held (2006, pp. 10–13).

[149] Held (2006, p. 10).

[150] Held (2006, p. 11).

[151] Noddings (1986, p. 4–5).

that all humans remember and long for. According to Noddings "It is the recognition of and longing for relatedness that form the foundation of our ethic, and the joy that accompanies fulfillment of our caring enhances our commitment to the ethical ideal that sustains us as one-caring."[152] In other words, Noddings's view of ethics is similar to Ubuntu's view, at least in as much as relatedness is its basis and objective.

Like Noddings, Ubuntu recognizes that care is universal to all humans since survival of human life from its tender and fragile beginnings, depends on care. The universality of care, according to Noddings can easily be found in the "caring attitude, that attitude which expresses our earliest memories of being cared for and our growing store of memories of both caring and being cared for." Thus, care is "universally accessible." Just as care requires a minimum of two people, so is personal goodness. Noddings contends that human beings are dependent on each other "even in the quest for personal goodness." This assertion makes personal goodness a joint venture with contingence to an-other. She states "How good I can be is partly a function of how you—the other—receive and respond to me. Whatever virtue I exercise is completed fulfilled, in you. The primary aim of all education must be nurturance of the ethical ideal."[153] Ubuntu could not agree with Noddings more when it comes to weighing morality and personal goodness by the "other," even if caring is natural to humans. Noddings indicates that caring is more feminine than masculine, thus she laments that "ethics has been discussed largely in the language of the father: in principles and propositions, in terms such as justification, fairness and justice." In her view, "principles and propositions" or quest for "justification, fairness and justice" are more masculine than they are feminine; more fatherly than they are motherly. In her view, "human caring" and "memory of caring and being cared for" are feminine than masculine; more motherly than they are fatherly. Consequently, she laments "The mother's voice has been silent.

Human caring and the memory of caring and being cared for, which I shall argue form the foundation of ethical response, have not received attention except as outcomes of ethical behavior."[154] Noddings argument implies that ethics based on principles and justice lacks an important aspect: empathy. She describes it by saying, "Apprehending the other's reality, feeling what he feels as nearly as possible, is the essential part of caring from the view of the one-caring."[155] Noddings position with regards to gendering ethics is not popular within Ubuntu culture. Ubuntu has been described as one which combines both justice and care.

Noddings then rejects the ethics of principle due to its having too many exceptions. Contrary to popular belief, she argues that ethics of principle is "ambiguous and unstable…too often principles function to separate us from each other." Since ethics of principle relate to universality Noddings rejects universability on the grounds that it wrongly overrides "Uniqueness of human encounters." In her view,

---

[152] Noddings (1986, p. 6).

[153] Noddings (1986, p. 6).

[154] Noddings, Nel. 2003. *Caring: A feminine approach to ethics and moral education*, 1. Berkeley and Los Angeles: University of California Press.

[155] Noddings (1986, p. 16).

in any ethical situation, too much depends on the subjective experience of the parties involved to appeal to objectivity and universality. Without enough explanation, however, Noddings declares that there is "a fundamental universality in our ethic, as there must be to escape relativism."[156] The grounds for the fundamental universality that Noddings claims in her brand of care, ethics have not been established except for the memories of care that enable people to crave for it. Noddings, therefore finds herself in a dilemma which forces her to admit the importance of principled approach to ethics which she tries to object. She gets out of the trap by admitting the need for some universality. Ubuntu ethic does not have the problem of disentanglement and prioritizing of care and justice. With Ubuntu human relationship, care and justice belong together.

3.3.1.2   Held on Care Ethics against Rational Traditional Ethics Principles

Held's concern is to provide a clear definition of care ethics against traditional justice based ethics. Ubuntu disagrees with Held's understanding of ethical care. According to Held one of the features that define ethics of care is the "compelling moral salience of attending to and meeting the needs of the particular others for whom we take responsibility."[157] Notably, here is Held's emphasis on particularity the care recipient and the willful assumption of responsibility by the care giver. In her view, care cannot be simply universalized or generalized. The second feature of ethics of care provided by Held is its holistic approach.

Thus, unlike the traditional ethics, "ethics of care values emotion rather than rejects it."[158] The third distinguishing feature of ethics of care is that "ethics of care rejects the view of the dominant moral theories that the more abstract the reasoning about a moral problem the better because the more likely to avoid bias and arbitrariness, the more nearly to achieve impartiality."[159] Even if Ubuntu favors concreteness, it also acknowledges the importance of universal principles. According to Held the approach embraced by ethics of care contradicts abstraction and embraces concreteness. The fourth distinguishing feature of ethics of care is that it seeks to distinguish and respect between that which is private from that which is public. The fifth distinguishing feature of ethics of care is its relational approach to moral problems. "While dominant moral theories tend to interpret moral problems as if they were conflicts between egoistic individual interests on the one hand, and universal moral principles on the other ... ethics of care, in contrast, focuses especially on the area between these extremes."[160] Thus, rather than polarize, ethics of care tend to unify and reconcile the parties and their perspectives in resolving moral problems.

---

[156] Noddings (1986, p. 5).

[157] Held (2006, p. 10).

[158] Held (2006, p. 10).

[159] Held (2006, p. 11).

[160] Held (2006, p. 12).

While it works against traditional ethics principles, Impartiality works for the ethics of care. The traditional approach tended to avoid bias in service of human equality and eliminate emotion and feeling from principle, mainly because they cannot easily be objectified and universalized. "An ethic of care focuses on attentiveness, trust, responsiveness to need, narrative nuance, and cultivating caring relations."[161] Thus ethics of care, at least from Held's perspective embraces bias and emotion based on each human being's uniqueness, uniqueness of human relationships, and, above all, uniqueness of each ethical situation.

While the claim of care ethicists carries with it some truth, it is hard to have plausible ethics devoid of objectivity, universality and principled. In an attempt to establish superiority of care over justice Held states, "Ethics of care usually works with a conception of persons as relational rather than as the self-sufficient independent individuals of the dominant moral theories."[162] The problem with her assertion is that it tends to undermine and override rational thought, which is a distinguishing feature of humanity.

In spite of Held's arguments against what she calls dominant moral theories, she does imply, unlike many other care ethicists, that justice may play a complementary role to ethics of care. Justice still has some role in ethics. It takes priority in some respective moral situations.[163] Generally, however care takes irreplaceable precedence over justice she posits that care is the deepest fundamental value since "there can be care without justice: there has historically been little justice in the family, but care and life have gone on without it. There can be no justice without care, however, for without care no child would survive and there would be no persons to respect." Although she establishes possible independence of justice from care, Held argues that care may function as the parameters within which justice should be sought. She states, "Care may thus provide the wider and deeper ethics within which justice should be sought."[164]This assertion, that care takes precedence over justice, is generally shared by Ubuntu philosophy.

Even though Held argues that care takes precedence over justice, her understanding of care remains too subjective between the caring and cared-for persons to be objectified or theorized. However, caring people in Held's view "are not seeking primarily to further their own *individual* interests; their interests are intertwined with the persons they care for."

Held holds that ethics of care is really applicable between people who are intimates or physically close with each other. She states that caring people "neither are they acting for the sake of *all others* or *humanity in general*; they seek instead to preserve or promote an actual human relation between themselves and *particular others*."[165] Probably because of this crippling dilemma Held had to admit that "the

---

[161] Held (2006, p. 15).

[162] Held (2006, p. 13).

[163] Held (2006, p. 17).

[164] Held (2006, p. 17).

[165] Held (2006, p. 12).

ethics of care may not itself provide adequate theoretical resources for dealing with issues of justice."[166]

In sum Held struggles to assert plausibility of ethics of care but she is realistic enough to realize the need for principled ethics, the need for universality and objectivity. While such needs have tended to trivialize human emotion, intimacy and care, they are nonetheless crucial in ethics. The experience-based approach of Ubuntu to morality is, in this case, more plausible than Held's perspective. Crucial for the survival of our human species as care may be, care cannot stand for itself independent of the traditional principles as stable, valid and plausible ethics. Ubuntu prioritizes care for intimates, other humans and environment in different degrees without underrating moral principles such as justice. It accepts humans' need for care from other humans; interdependence, and the limits of both subjectivity and objectivity. Such limits indicate the importance of both care and justice or principle.

3.3.1.3 Slote on the Role of Empathy in Care Ethics

Slote argues that ethics of care is not new. In his view, it has been there though different in wording, use and objective. Slote states that ethics of care falls within "the ethical tradition known as moral sentimentalism." One of the philosophers of moral sentimentalism is David Hume. Notably, Hume never used the name ethics of care but used words such as "benevolence, compassion and sympathy."[167] Using psychological data, Slote argues that ethics of care is based on human ability and need to empathize. As such, it is universal to human beings. Slote contends that care ethicists either have not, or have failed to provide a robust theory of moral education supporting their kind of ethics.

Basing his reasoning on Martin Hoffman's work, Slote argues that "empathy is central both to moral education and to moral development." Slote, therefore, challenges care ethicists to "pay more attention to the psychological literature on empathy and moral development."[168] Slote's argument entails that ethics of care specifically, and all ethics generally, is based on human relationship and the ability to empathize. Empathy, he contends, "is relevant to right and wrong." Failure to empathize would, in other words render ethics of care meaningless.[169] Unlike most care ethicists such as Noddings, Slote does not base the foundation of ethics of care on memory but on a human ability to empathize. This foundation aligns care ethics with Hume's theory of benevolence.

Like Held and a number of other care ethicists, Slote argues that since care is based on empathy, it is generally more relevant to those geographically or otherwise closely related to the subject. Moral obligation to care is equally subject to both geographical and relational proximity. The difference that sets him apart is basing

---

[166] Held (2006, p. 17).
[167] Slote (2007, p. 4).
[168] Slote (2007, p. 4).
[169] Slote (2007, pp. 5–8).

his theory of care on empathy. However, empathy and care should not be limited to the close others, although it is increased by personal relationships and geographical proximity.[170] In slote's view, empathy cannot be excluded from morality.[171] Slote, complies with most care ethicists that care is of fundamental importance to morality, even if most care ethicists don't relate ethics of care with empathy.

Slote argues in favor of ethics of care as capable of promoting respect to people and their self-determination. This is clearly stated in his own words "I believe a sentimentalist ethics of care can, in fact, ground respect, and respect for autonomy, in its own terms."[172] In his view, therefore, ethics of care can complement traditional principle-based ethics. Slote implies that respect is based on, flow from or motivated by empathy when he states, "respect for individuals can be unpacked in terms of empathy…one shows respect for someone if, and only if, one exhibits appropriate empathic concern for them in one's dealings with them."[173] Thus, unlike many care ethicists, Slote does not argue that there is necessary conflict between what he calls sentimentalist ethics or ethics of care and the traditional ethics. The two types of ethics not only complement each other, they need each other. People need both types of ethics. Slote raises a valid major problem with ethics of care. That is its tendency to self-destruct because of its auto-contradiction. Slote observes that some ethicists of care have argued that care ethics "is more appropriate to women than to men," while, at the same time some care ethicists and other thinkers have "claimed that care ethics works against the goals of feminism by recommending the very attitudes and activities that have kept women subordinate to men throughout the ages."

Slote notes that these two schools of thought within ethics of care "are in some tension with one another, but either of them could lead one to conclude that care ethics cannot function, or function well, as a morality governing both men and women." However, Slote believes that "a fully elaborated ethics of care has the potential to function in a comprehensive and satisfying way as a truly human morality."[174] Thus, ethics of care, though potentially possible universally, it bears within itself seeds of its own destruction.

Slote's approach to care ethics has a lot in common with Ubuntu's. Care is not exclusive of principles, especially principles of justice. Both care or empathy and justice or principles are equally needed for a happy society. A comprehensive ethic cannot choose between principles and care, or between rationality and emotion. To be credible and plausible, an ethic needs to be inclusive both of such trends and people, regardless their gender. One of the greatest credits about Ubuntu is its refusal to over classify and compartmentalize. It is simply holistic in approach. Empathy, care, justice, principles and emotions belong to what constitutes a humane person. Moral maturity consists of all those human values.

---

[170] Slote (2007, p. 5).

[171] Slote (2007, p. 127).

[172] Slote (2007, p. 56).

[173] Slote (2007, p. 57).

[174] Slote (2007, p. 8).

3.3.2 Care Universalized

There are two schools of thought among ethicists of care: those who argue that real care exists only between loved ones and those immediately related,[175] and those who contend that care can be, and ought to be, universalized.[176] Noddings, for example, roots care ethics in the "attitude which expresses our earliest memories of being cared for and our growing store of memories of both caring and being cared for." She contends that care is "universally accessible."[177] All people have experiences of being cared for, and most have experiences of caring for others. Hence they intuitively recognize care as good. Everyone implicitly acknowledges the morality of caring relations even if only among family or friends. To reject care on principle is to reject the basic conditions of human development and sociability.[178] Some feminist scholars have criticized and resisted any attempt to limit care to private relations and contexts but have failed to show how care ethics can be translated into applicable consistent and robust universal moral theory.[179]

According to Noddings and Held, in a strict sense, no institution or nation can be ethical. Organizations cannot meet one another as one caring or as one trying to care. Organizations "can only capture in general terms what particular one's caring would like to have done in well-described situations."[180] A perspective like this would imply that general legal rules and policies could do violence to the particular and variable needs of individuals. Gilligan established care as a major new perspective in the discourse on contemporary ethics. However, she did not establish it as the sole valid approach; rather, she indicated that her theory is only supplementary to theories of justice.[181]

The problematic conflict of universalization of care between the two schools of thought generates the question of validity of care as ethics. If care is completely bound in particularity and uniqueness of each particular human relationship there cannot be a discourse on ethics of care. If care can be generalized it needs to have robust principles and rules to be a credible ethic. This problem of ethics of care is shared by Ubuntu ethics. While care in Ubuntu should be the norm, it is both too universal and too dependent on context that it is caught in the same dilemma as care ethics.

---

[175] Slote (2007, pp. 9–25).

[176] Tronto (1993).

[177] Noddings (1986, p. 5).

[178] Noddings (1986, p 49, pp. 79–81).

[179] Tronto (1993, pp. 87–91).

[180] Noddings (1986, p. 103).

[181] Gilligan (1982, pp. 173–174).

3.3.2.1 Robinson on the Problem of Globalization of Care

Robinson's perspective on ethics of care is radically different from that of Gilligan, Held, Tronto and others. According to her, "Moral reasoning and ethical enquiry which take care as its starting point do not seek to construct a moral theory at all." She asserts that ethics of care is "a collection of perceptive, imaginative, appreciative and expressive skills and capacities which put and keep us in unimpeded contact with the realities of ourselves and specific others." To exemplify her assertion Robinson writes, "A critical ethics of care does not seek to arrive at an account of moral philosophy which presents a justification for action dependent on the application of principles and rules."[182] Thus Robinson does agree with majority of care ethicists that ethics of care is not meant for evaluation of any single human act using some set principles or theories. It does not specifically prescribe what should or should not be done in an ethical situation.

According to Robinson, an ethic of care is relation-specific. Since it is "neither categorical nor universal-prescriptive; it does not demand that we 'care' wholly, and equally, about all individuals at all times in all places, nor does it regard a moral response as an act of pure will or judgment." Ethics of care is based on, and evaluates actual or possible relations and the "capacity of those agents to learn how to listen and respond to the needs of others." Therefore ethics of care is implicitly different from the traditional principled ethics in its objectives and methodology. While ethics of principle evaluate human action as right or wrong, ethics of care evaluates human dispositions to, and quality of relationships. According to Robinson ethics of care cannot "provide an answer to the question that plagues normative theorists of international relations: how to arrive at global or universal norms/values in a world of particular, competing, and often incommensurable value systems."[183]

Viewing ethics of care from Robinson's perspective makes it hard to effectively explore it, especially because it is so limited to the subjective experience of those locked in a personal relationship. However, Robinson posits that ethics of care is credible because it instills the sense of *oughtness*. She writes, for example that "Those who are powerful have a responsibility to approach moral problems by looking carefully at where, why, and how the structures of existing social and personal relations have led to exclusion and marginalization, as well as at how attachments may have degenerated or broken down so as to cause suffering."[184]

The problem, however, lies in the evaluation of this obligation since there are really no objective principles to guide such evaluation. Robinson further argues that "Moral response is not a rational act of will, but an ability to focus attention on another and to recognize the other as real. Such recognition is neither natural nor presocial, but rather something that emerges out of connections and attachments."[185]

---

[182] Robinson (1999, pp. 39–40).

[183] Robinson (1999, p. 40).

[184] Robinson (1999, p. 46).

[185] Robinson (1999, p. 46).

Her attempt to at least limit or at most eliminate human rationality and volition from morality makes it very hard to establish credibility of ethics of care as it makes it less of an academic discourse.

One of the significant contributions of Robinson to ethics of care is her recognition and envisioning of its global dimension. While Robinson like Held argues that ethics of care is really between intimates and those close to each other physically, she contends that it should inspire global care between nations and institutions. In her view, this recognition comprises of, at the very basic stage, recognition of others as real and needy.[186] She challenges ethics of care to be global and to evaluate unethical global structures which cause or support exclusion and address such structures. She boldly asserts that "Any approach to ethics which claims to address the moral problems of international relations cannot overlook the structural causes of patterns of moral inclusion and exclusion on a global scale."

In her view, there is need to reflect critically on structures which compromise individual person's reality and uniqueness. She sees a need to address structures which by "'community-making', and hence exclusion, serve to undermine the ability of moral agents to identify and understand others as 'real' individuals—with real, special unique lives."[187] Thus Robinson's perspective on global ethics of care is basically inclusion. Ethics of care's ideal on international level should be inclusion of the entire human family and elimination of the problem of human marginalization.

Robinson observes that there is a trend towards, creation of global human family with common identity.[188] However, globalization in Robinson's view is structurally flawed if it does not address the "exclusionary social practices and structures in the contemporary global system: how boundaries are constructed, how 'difference' is assigned, and how moral and social exclusion is legitimized."[189] In other words the process of creation of global community is at the same time excluding and marginalizing some members of the community it seeks to create. Robinson wages a war against what she termed "institutionalization of exclusion."

She argues that "Understanding obstacles to moral responsiveness among distant strangers simply in terms of ignorance, egoism, or individual prejudice obscures the 'institutionalization of exclusion' which occurs not only within political communities but between them."[190] Robinson believes that critical ethics of care may be the solution needed within the international relations theory. She states, that "A critical ethics of care could eclipse the quintessential problem of international relations theory: resolving the conflict between our 'egoistic' roles and duties as citizens and our 'altruistic' roles and duties as human beings."[191]

Robinson's Ethics of care resembles Ubuntu in many ways, especially in its pursuit of human beings' common identity, creation of global human community,

---

[186] Robinson (1999, pp. 48–50).
[187] Robinson (1999, p. 47).
[188] Robinson (1999, p. 91).
[189] Robinson (1999, p. 131).
[190] Robinson (1999, p. 131).
[191] Robinson (1999, p. 50).

recognition of other people's humanness regardless their social, economic or racial difference and its emphasis on individual's duty to pay attention to the needs of other humans.

3.3.2.2 Sevenhuijsen On the Moral Dilemma of Feminism in Ethics of Care

Sevenhuijsen writing implies that ethics of care is going through an identity crisis. One of the problems causing the crisis is feminism. "The feminist discussion about an ethics of care is too heavily weighted towards questions of identity rather than questions of agency and morality."[192] Studies by Gilligan and other feminist ethics of care have tended to ground justification of their feminist theories on psychological and cognitive developmental theories like Kohlberg's. The result of such researches creates and exaggerates the chasm between the two genders. A good example is Chodorow. In distinguishing a masculine conception of care from a feminine one Chodorow writes,

> Girls come to experience themselves as continuous with others; their experience of self contains more flexible or permeable ego boundaries. Boys come to define themselves as more separate and distinct, with a greater sense of rigid ego-boundaries and differentiation. The basic feminine sense of self is connected with the world; the basic masculine sense of self is separate."[193]

When translated into ethics, Chodorow's assertion may demand different standards be adopted for the two genders. Even though there may be truth in Chodorow's assertion, its exaggeration is counterproductive.

There is an implied tension and conflict between morality and feminism. Sevenhuijsen points out the problem. She states that "The relationship between feminism and morality has recently been one of unease and suspicion. Many modern feminists see morality as one of the phenomena from which women should be liberated, and they easily associate it with paternalism, restrictive regulation of women's lives, and conservatism."[194] Although ethics should ideally be contra oppression, exploitation and marginalization of any person or any group of persons, there has been a feeling among feminists that morality itself has been exploitative, marginalizing, exclusivist or ignoring women. In other words, ethics has not treated the female gender as equal to the male gender. Thus Sevenhuijsen writes that "It is neither an easy nor an inviting proposition for feminism to relinquish the norm of equality. The idea that men and women do not differ systematically in their capacities is an indispensable element in feminism whose principle objective is the fair treatment of men and women."[195] However, feminist ethicists like Gilligan, who have either written or implied that ethics of care is feminist, do indirectly imply that there is inequality between the two genders, even in its view of morality. Ethics of care would,

---

[192] Sevenhuijsen (1998, p. 25).

[193] Chodorow (1978, p. 169).

[194] Sevenhuijsen (1998, p. 36).

[195] Sevenhuijsen (1998, p. 42).

therefore, be self-destructive by demanding gender equality while, at the same time, affirming inequality as a given.

Sevenhuijsen argues for the importance of difference. She states that "it is because people are treated differently that they depart from their natural sameness."[196] Sevenhuijsen observes that "the public debate about gender issues repeatedly ends up in a vicious circle;"[197] often times such debates do not consider their own relevance. They are often counterproductive even as they escalate in intensity. Sevenhuijsen argues that "a universalist ethics implies that women should conceive of themselves as little as possible in terms of sexual difference."

They should immerse themselves in the world community. Equality in outcome is not always ethical. Sevenhuijsen states that "the principle that equality in results should be the touchstone for politics and policy has the effect of marginalizing other moral questions, such as the question of how oppression, violence, vulnerability and plurality should be dealt with, or how quality of life can be improved."[198] The plausibility of Sevenhuijsen's argument consists in its realistic observation which in many ways conforms to Ubuntu worldview. Difference and diversity, even in gender, is constructive and productive for the human global community.

Sevenhuijsen addresses a centuries year old Platonic and Cartesian influence on the western culture which shapes not only the culture but also the thinking and human treatment of each other. This influence works on difference:

> Under the influence of philosophers such as Plato and Descartes, Western culture has been permeated by thinking in terms of oppositions or mutually exclusive opposites. The most important of these are the oppositions between culture and nature, reason and emotion, spirit and body, rational and material, rational and emotional, fatherhood and motherhood, freedom and necessity, self and other, inner and outer, universality and particularity, public and private, Western and Eastern, civilization and primitiveness, and masculine and feminine, independence and dependence... True personhood is defined as the dominant norm and the 'other' can have no subjectivity of its own; this process is termed the 'objectifying of otherness.[199]

Objectifying of otherness is one of the major differences between the Ubuntu worldview and the Western worldview. Ubuntu philosophy encourages recognition of the subjectivity of the other, even as the other becomes an object of one's thought or action. Within Ubuntu culture otherness is exalted because it is necessary for selfhood. It plays such an important role that the maxim "I am because you are" is the only one which briefly but exhaustively explains it. In the culture of Ubuntu, exclusion whether of self or other is equivalent of homicide. Inclusion is the ideal. Western culture remains trapped in the 'subject—object' dilemma.

Ubuntu agrees with Sevenhuijsen regarding gendering ethics. Ethics of care ought not be used as feminist ethics. The trend of reducing care ethics into a kind of particular specialized ethics for the female gender has been, and still is the

---

[196] Sevenhuijsen (1998, p. 42).
[197] Sevenhuijsen (1998, p. 42, 43).
[198] Sevenhuijsen (1998, p. 43).
[199] Sevenhuijsen (1998, p. 47).

most destructive of its meaning, significance and implication. Ethics should not be confused with sexual/gender or developmental psychology. Ubuntu has a much better grip on the meaning of ethics of care than those who work hard to limit it to the feminine gender. In the following section Clement deals with the problem of gendering ethics of care.

3.3.2.3   Clement Liberating Ethics of Care from Feminism for Virtue Ethics and Justice

The uniqueness of Clement's work consists of her careful critique of ethics of care as it is perceived by other care ethicists. She rejects the three most popular trends of ethics of care: "the celebration of the ethic of care as a feminine ethic, the assimilation of the ethic of care to a justice perspective and the rejection of the ethic of care from a feminist perspective." She does support the view that ethics of care should be virtue ethics, or at least part of virtue ethics as viewed by Aristotle. She explores ethics of care on par with ethics of justice without letting either of them undermine the other.[200] Clement explores two features of ethics of care which she considers to be central. The first feature is Justice and the second feature is "ethic of care's status as a personal ethic."

Clement observes that "feminine advocates of the ethic of care argue that autonomy is an individualistic value that the ethic of care rejects in favor of relational virtues. However, its feminist critics argue that because the ethic of care compromises a care giver's autonomy, it fails by feminist standards." With regards to the second feature of ethics of care which is equally controversial, "the ethic of care's status as a personal ethic, it is appropriate for our relations with family, friends or those otherwise close to us, such as students." Thus ethics of care would seem to complement ethics of principle/justice rather than rival it.

Grim explains the relationship between autonomy and relation to underline their interdependence. He argues that women's experience of connectedness may as well work against the very ethic of care which assumes and emphasizes empathy and connectedness. Grimshaw explains, "If I see myself as 'indistinct' from you, or you as not having your own being that is not merged with mine, then I cannot preserve a real sense of your well-being as opposed to mine." There is need to respect boundaries which leave room for personal autonomy. Moreover, as Grimshaw observes, "Care and understanding require the sort of distance that is needed in order not to see the other as a projection of the self, or self as a continuation of the other."[201]

Complementarity and a sense of balance between autonomy and relationship are essential. For feminist care ethicists, ethics of care "gives personal relations the moral attention they deserve, correcting the ethic of justice's view of personal relations as morally insignificant in comparison to public relations. Conversely, its

---

[200] Clement (1996, pp. 2–4).

[201] Grimshaw (1986, p. 183).

critics argue that a feminist ethic must not be limited to personal relations, and must include a concern for social justice."[202]

The reconciliatory role of Clement's writing gives both ethics of care and ethics of justice recognition of its neediness of the other. The two ethics are not mutually exclusive but mutually complementary. Clement argues that "the conflicts between care and justice orientations need not lead us to accept one at the expense of the other; indeed, these conflicts can help us distinguish between better and worse versions of each ethic."[203] Care and justice have been viewed as mutually exclusive alternatives to each other since "they are understood as conflicting ethics, each with its own ontology, method, and priorities, committed to mutually exclusive values and best suited to different kinds of situations." The two ethics are generally distinguished in three ways:

1. the ethic of justice takes an abstract approach, while the ethic of care takes a contextual approach;
2. the ethic of justice begins with an assumption of human separateness, while the ethic of care begins with an assumption of human connectedness; and
3. the ethic of justice has some form of equality as a priority, while the ethic of care has the maintenance of relationships as a priority.

These features in turn are generally taken to result in conflicting evaluations of autonomy and a division of labor between the two ethics along public/private lines.[204]

The rivalry between ethics of care and ethics of justice escalates as one seeks to embrace one while attempting to disqualify and reject the other. This extremist tendency is unfortunate because there is a genuine and necessary relationship between care and autonomy. She posits that "feminist ethic of care must allow for its adherents' autonomy."[205] Since autonomy is a "moral competence that has both personal and social dimensions," argues clement, "the commonly held view that care and autonomy are mutually exclusive arises because of the excessively individualistic and excessively social conceptions of the self that accompany the ideal types of justice and care." Ideally care and autonomy should be interdependent.[206]

Equally misconceived and misjudged is the relationship between abstract and concrete and between reason and emotion. Clement states observes that "From the justice perspective, feelings are seen as threatening the universality demanded of moral judgment, and thus we should seek to abstract from our particular feelings and focus on universal principles to be properly moral."[207] Abstracting from particular individual feelings in order to focus on universal principles does injustice to the validity and relevance of the principles.

---

[202] Clement (1996, p. 7).
[203] Clement (1996, p. 7).
[204] Clement (1996, p. 11).
[205] Clement (1996, p. 8).
[206] Clement (1996, p. 8).
[207] Clement (1996, p. 13).

Similarly, ignoring the principles for the sake of the particular concrete emotions within a particular ethical situation renders that very situation impossible to evaluate ethically, for lack of criteria. Knowing the better or worse versions of each of the two ethics would enrich bioethics and ethics tremendously. Although Clement's perspective is similar to Ubuntu worldview, the striking similarity is in her view and role of context. Clement warns that "just as it is a mistake to ignore care's social context, it is also a mistake to *reduce* the ethic of care to the distorted ways it is often practiced."[208] While ethics of care cannot be reduced to mere context of an ethical situation, ethics cannot be independent of context. To disentangle context from ethics is equivalent to rendering ethics devoid of substance.

Similarly, the role of human equality and attachment in ethics of care should not be taken for granted. Clement contends that "All human relationships, public and private, can be characterized *both* in terms of equality and in terms of attachment, and ... both inequality and detachment constitute grounds for moral concern." In themselves the concepts of equality and attachment may demonstrate the interdependence between ethics of care and ethics of justice. Clement states, "Since everyone is vulnerable both to oppression and to abandonment, two moral visions—one of justice and one of care—recur in human experience.

The moral injunctions, not to act unfairly toward others, and not to turn away from someone in need, capture these different concerns."[209] In sum, the attempt to separate care from justice and to prioritize one type of ethics over the other is not only wrong and unfair, it is counterproductive to both ethics of care and ethics of justice. The Ubuntu approach which prioritizes and recognizes care's precedence over justice while, at the same time, recognizing the indispensable role of justice in care and human social life is the most relevant approach.

3.4   Conclusion

Ubuntu worldview recognizes human cognitive and moral development and proactively facilitates it by continuous initiation rites from instinctive self-centered struggle for survival mode of an infant to a mature autonomous but also social and moral person. Although not in the same organized and systematized manner as elaborated by Kohlberg and Gilligan, Ubuntu has the basic understanding of human moral development. One of the differences between Ubuntu and modern theoretical theories is that Ubuntu's method is based on centuries of lived observation and experimentation, has been passed on verbally, and is always praxis based. By progressive staged initiation a child is helped to grow into the stage where a child can say "*I am because you are*" and the society knows then that the child is really a mature person who can be left to act all the time for the self and the society.

---

[208] Clement (1996, p. 6).
[209] Gilligan (1987, p. 20).

Without creating conflict between personal autonomy and human need for relationships (interdependence), Ubuntu recognizes the role and boundaries of self-determination (autonomy) relative to relationships and societal boundaries into personal autonomy. The society helps every individual grow into his unique actualization as person while not letting him slip into the dangerous distance that disconnects him from the rest of the society. There is no conflict between human need for both care and justice in Ubuntu. There is not even a separation between the two. Justice and care are concomitant and concurrent.

They are perceived as two sides of the same coin. Hence the conflict between them is not even perceived. Human emotion and feelings are accepted as real and addressed accordingly by the society for the good of the individual involved and for the good of all other members of the society since there is the necessary interconnection between members of the society which allows all to empathize, and therefore, act on behalf of, and for each other. Physical, psychological, emotional and moral difference between male and female genders is accepted as a blessing since complementarity is the order of life. While basic human dignity equality is *sine qua non*, uniqueness and difference is embraced as richness for all.

Having enlightened Ubuntu's human and societal perspective by ethics of care, the following section will explore Ubuntu's perspective on global human society and its relationship with the biosphere and the inorganic part of the cosmos as the necessary context of human life and society. The next section will rely heavily on the UNESCO Universal Declaration on Bioethics and Human Rights as it relied on ethics of care in this section.

Chapter 4
UNESCO Declaration: Enlightening the Cosmic Context of Global Bioethics

One of the most important components of the culture of Ubuntu is its respect for the essential cosmic/global context. The meaning of this context can be enlightened by considering the UNESCO Universal Declaration on Bioethics and Human Rights. One of the major components of the UNESCO Universal Declaration on Bioethics and Human Rights concerns justice. The ethical debate on human rights respects the universal primacy of the human person within the parameters of the principle of justice. Another major component of the UNESCO Universal Declaration on Bioethics and Human rights is based on diversity.

The debate on ethical responsibility must respect cultural and racial diversity within a global context. Another important component of the UNESCO Universal Declaration on Bioethics and Human Rights is respect for the biosphere. Respecting diversity includes respect for the biosphere as the cosmic context for discourse on ethical responsibility. This chapter explores all three components of UNESCO Declaration on Bioethics and Human Rights to enlighten Ubuntu's aspect of universal/global context. Before elaborating on the major themes of comparison between Ubuntu and the UNESCO's Universal Declaration on Bioethics and Human Rights section one gives a brief analysis of the articles of the declaration from Ubuntu perspective.

4.1 UNESCO Declaration on Bioethics and Human Rights from Ubuntu Perspective

The UNESCO Declaration on Bioethics and Human Rights is to a great extent formalization and systematization of the ideals of Ubuntu. Many scholars who understand indigenous cultures and their objectives, especially with regards to ethics and morality, realize that most articles of the UNESCO Declaration on Bioethics and Human Rights verbalize the content and ideals of such indigenous cultures. This section demonstrates the similarities between Ubuntu and UNESCO's Declaration on Bioethics and Human Rights. Needless to mention, Ubuntu represents many indigenous cultures.

L. T. Chuwa, *African Indigenous Ethics in Global Bioethics,* Advancing Global Bioethics,   137
DOI 10.1007/978-94-017-8625-6_4, © Springer Science+Business Media Dordrecht 2014

4.1.1   Articles Regulating Societal/National, and Global Behavior

The UNESCO Declaration on Bioethics and Human Rights is not only about personal ethics. It is an official text of international and global bioethics. Even though its effect covers all individuals, the code transcends individuals to deal with their socio-geographical contexts. Much as it seeks to safeguard the good of the human race presently alive, the Declaration also transcends the present generation to protect the common good of humanity both now and in the future. The articles of the UNESCO Declaration on Bioethics and Human Rights helps one see how Ubuntu as a very simplified pragmatic philosophy of life aimed at, and worked on the same objectives and ideals.

4.1.1.1   Scope and Aims

Article number 1 section one of the Declaration, "addresses issues related to medicine, life sciences and associated technologies as applied to human beings, taking into account their social, legal and environmental dimensions." Section two of article 1 states that the "Declaration is addressed to States. As appropriate and relevant, it also provides guidance to decisions or practices of individuals, groups, communities, institutions and corporations, public and private."[1] Although Ubuntu is neither formalized nor systematic, it applies both to human beings as individuals and as a species. The treatment an individual or society gives to another individual or society define the subject. Life is relationship. According to Ubuntu, life devoid of relationship whatsoever is void. So many sayings and proverbs remind the society the importance of quality human relationship. A relationship does not only define the parties involved in it, it defines existence itself. Article number 26 cautions that the Declaration should be treated holistically as one document since its principles are interrelated and complementary to one another. It goes, "This Declaration is to be understood as a whole and the principles are to be understood as complementary and interrelated. Each principle is to be considered in the context of the other principles, as appropriate and relevant in the circumstances."[2] This methodological article of the Declaration is so much similar to Ubuntu methodology. Ubuntu wisdom, guidance, regulations and ideals are all summarized and contained in this maxim: *a human being is human because of other human beings*. One's actions should reciprocate the goodness that he/she has received from others/community. Personal actions should contribute to community's project of creating and fostering individual and communal life. Specifics and details are not as important. Ubuntu is holistic in approach.

Article 27 underline the fact that state laws should "be consistent with international human rights law." It elaborates, "If the application of the principles of this Declaration is to be limited, it should be by law, including laws in the interests of public safety, for the investigation, detection and prosecution of criminal offences,

---

[1] Andorno (2009b, p. 67).

[2] Andorno (2009b, p. 327).

for the protection of public health or for the protection of the rights and freedoms of others."[3] Like the Declaration Ubuntu applies universally to all human beings regardless their location on the globe. However, specific societies/communities may discipline or even execute a constituent who threatens either the life of the community as a whole or other lives.

Article number 28 denies "acts contrary to human rights." The Declaration explains, "Fundamental freedoms and human dignity Nothing in this Declaration may be interpreted as implying for any State, group or person any claim to engage in any activity or to perform any act contrary to human rights, fundamental freedoms and human dignity."[4] Some scholars from the Western hemisphere misjudge Ubuntu as a kind of communistic dictatorship which does not care for individual rights, freedoms and dignity. The truth is, Ubuntu does care for all those individual rights and entitlements just as much as any other modern societal system. The difference is, Ubuntu cares for an individual necessarily within the matrix of society. Ubuntu fails to find individual rights outside the society or community because, in its view, no individual can survive outside the society.

Article 2 of the Declaration states the aims of the Declaration, which are:

a. To provide a universal framework of principles and procedures to guide States in the formulation of their legislation, policies or other instruments in the field of bioethics;
b. To guide the actions of individuals, groups, communities, institutions and corporations, public and private;
c. To promote respect for human dignity and protect human rights, by ensuring respect for the life of human beings, and fundamental freedoms, consistent with international human rights law;
d. To recognize the importance of freedom of scientific research and the benefits derived from scientific and technological developments, while stressing the need for such research and developments to occur within the framework of ethical principles set out in this Declaration and to respect human dignity, human rights and fundamental freedoms;
e. To foster multidisciplinary and pluralistic dialogue about bioethical issues between all stakeholders and within society as a whole;
f. To promote equitable access to medical, scientific and technological developments as well as the greatest possible flow and the rapid sharing of knowledge concerning those developments and the sharing of benefits, with particular attention to the needs of developing countries;
g. To safeguard and promote the interests of the present and future generations;
h. To underline the importance of biodiversity and its conservation as a common concern of humankind.[5]

---

[3] Andorno (2009b, p. 334).
[4] Andorno (2009b, p. 343).
[5] Andorno (2009b, p. 81).

As it has been mentioned above, Ubuntu is not formal but its objectives for the individual and universal good of the human beings of this generation and of future generations is clear and indisputable.

Article 16's objective is protection of future generations. The article aims at regulating "the impact of life sciences on future generations, including on their genetic constitution."[6] According to Ubuntu all life is sacred and it actually belongs to God. Human beings may not temper with life. It is the obligation of human beings to protect, nurture and cherish life as it comes from God through nature. In other words, Ubuntu philosophy would not condone taking risks with lives of future generations whether human, animate or vegetative.

Article 17 aims at protecting "the environment, the biosphere and biodiversity. The article states:

Due regard is to be given to the interconnection between human beings and other forms of life, to the importance of appropriate access and utilization of biological and genetic resources, to respect for traditional knowledge and to the role of human beings in the protection of the environment, the biosphere and biodiversity.[7]

Ubuntu hold nature as sacred, especially because of the role it plays to human life. As individual human beings cannot realistically be separated from the universal human society, so is the human species from nature. Bujo states, "African ethics treats the dignity of the human person as including the dignity of the entire creation, so that the cosmic dimension is one of its basic components."[8] This perspective underlines ethical conduct may be "based on the individual but is realized primarily by means of a relational network that is equally anthropocentric, cosmic, and theocentric."[9]

4.1.1.2   Ideals and Values Protected

Article number 10 states, "The fundamental equality of all human beings in dignity and rights is to be respected so that they are treated justly and equitably."[10]

The very kernel of the essence of Ubuntu philosophy is acknowledgement of basic human equality which must not only be recognized but which must be protected and respected. The Statement: "a human being is a human being because of other human beings" does not only reveal human symbiosis and mutuality but also human basic equality. This recognition implies and obliges the ethical principles of justice, beneficence, nonmaleficence and solidarity.

Related to article number 10, is article number 11 which forbid discrimination. It states, "No individual or group should be discriminated against or stigmatized on any grounds, in violation of human dignity, human rights and fundamental freedoms."[11]

---

[6] Andorno (2009b, p. 243).

[7] Andorno (2009b, p. 248).

[8] Bujo (2001, p. 2).

[9] Bujo (2001, p. 2).

[10] Andorno (2009b, p. 173).

[11] Andorno (2009b, p. 187).

Ubuntu has strong regulations that guide and guard the society against discrimination. Not even war captives would be discriminated upon. Instead, they would be adapted as members of the society. Orphans would neither be allowed to feel nor know that their biological parent/parents were dead. They would naturally be adopted by uncles or aunties. Children belonged to the entire society. Strangers would be welcomed, fed and accommodated. People of other ethnicities would be made to feel at home. Unlike the modern Western tendency, Ubuntu did not verbalize much about the seriousness of discrimination, it resisted it vehemently.

Also closely related to article number 10 and 11 is article number 12 which recognize cultural diversity and pluralism. The article urges for respect for diversity and pluralism but warns about the limits of cultural pluralism and diversity. It states,

"The importance of cultural diversity and pluralism should be given due regard.

However, such considerations are not to be invoked to infringe upon human dignity, human rights and fundamental freedoms, nor upon the principles set out in this Declaration, nor to limit their scope."[12] Among other values, Ubuntu is based on the recognition, not only of human essential equality, but also of human plurality and diversity. Ubuntu cherishes plurality and diversity as richness. Humans flourish on the otherness of others. In other words, it is human plurality and diversity that enrich each member of the society. Such diversity extends from personal to societal or national.

Article 13 is related to article 12. It states, "Solidarity among human beings and international cooperation towards that end are to be encouraged."[13] Just as a baby cannot make it by itself right after it is born, just as it needs other people to help it get gradually more independent, so does any individual remain in need of others/community for his/her self-actualization. For Ubuntu, human growth and development is a continuum that goes on from the womb into the society. It is within the society that one continually finds/realizes oneself. The deeper one relates with the society the more mature that person may become. Personal rights have to be enjoyed within the society because without the society the person does not exist. This principle of Ubuntu applies also for national states. Relationship and mutuality is crucial for human prosperity.

Article 14 is on social responsibility and health. It states:

1. The promotion of health and social development for their people is a central purpose of governments that all sectors of society share.
2. Taking into account that the enjoyment of the highest attainable standard of health is one of the fundamental rights of every human being without distinction of race, religion, political belief, economic or social condition, progress in science and technology should advance:

   a. Access to quality health care and essential medicines, especially for the health of women and children, because health is essential to life itself and must be considered to be a social and human good;

---

[12] Andorno (2009b, p. 199).

[13] Andorno (2009b, p. 211).

   b. Access to adequate nutrition and water;
   c. Improvement of living conditions and the environment;
   d. Elimination of the marginalization and the exclusion of persons on the basis of any grounds;
   e. Reduction of poverty and illiteracy.[14]

Relative to the wealth of the community/society, Ubuntu would unanimously and naturally set a poverty line below which no member of the society should be allowed to fall. In case of sickness or any condition that threaten or compromise human life, each member of the society would bring in his best contribution to save life regardless of the merits of the victim.

Article number 15 is based on distribution. It states,

1. Benefits resulting from any scientific research and its applications should be shared with society as a whole and within the international community, in particular with developing countries. In giving effect to this principle, benefits may take any of the following forms:

   a. Special and sustainable assistance to, and acknowledgement of, the persons and groups that have taken part in the research;
   b. Access to quality health care;
   c. Provision of new diagnostic and therapeutic modalities or products stemming from research;
   d. Support for health services;
   e. Access to scientific and technological knowledge;
   f. Capacity-building facilities for research purposes;
   g. Other forms of benefit consistent with the principles set out in this Declaration.[15]

African society's reverence for life would never allow human life to be used in any way as a means to another end, even if that other end is another human life. It was very much in line with Aristotelian teleology. All human beings ultimately crave happiness or happy life. However, individual happiness is not the ultimate end since, as Aristotle noted, "For a while the good of an individual is a desirable thing, what is good for a people or for cities is a nobler and more godlike thing."[16] In other words, the entire society is ultimately invested in the happiness of the society, which, in turn, is shared by the constituents of that society.

4.1.1.3   Implementation

Article number 19 of the UNESCO Declaration on Bioethics and Human Rights recommends establishment, promotion and support of "Independent, multidisci-

---

[14] Andorno (2009b, p. 218).
[15] Andorno (2009b, p. 231).
[16] Aristotle (2000, p. 4).

plinary and pluralist ethics committees." The Declaration explains the functions of such ethics committees as to;

a. Assess the relevant ethical, legal, scientific and social issues related to research projects involving human beings;
b. Provide advice on ethical problems in clinical settings;
c. Assess scientific and technological developments, formulate recommendations and contribute to the preparation of guidelines on issues within the scope of this Declaration;
d. Foster debate, education and public awareness of, and engagement in, bioethics.[17]

Traditional African society was organized partially according to the principle of subsidiarity. Division of labor, usually according to personal or group's capabilities and talents was a *modus operandi*; specialized details like formation of ethics committees were inexistent. However, the functions of ethics committees would naturally be performed either by elders, or medicine men/women or chiefs and their councils.

Article number 20 states, "Appropriate assessment and adequate management of risk related to medicine, life sciences and associated technologies should be promoted."[18] In the traditional African society risk assessment and the balance between beneficence and nonmaleficence was basically the function of medicine men/women. However, when it was evident that a member of the society was actively dying, postponement of death or prolongation of the process of dying was not considered ethical.

Ubuntu believe in the eschatological life hereafter. For an actively dying person, risks would be taken that would save the life of the ill member of the society, otherwise the sick person would be initiated into the world of the *living-dead*, using the words of Mbiti. It is on these grounds Bujo raises the controversial ethical question: "Is it not an offence to human dignity to prolong life by artificial means when only a vegetative life is possible, or when the inevitable death can only be postponed for a few hours or days?"[19]

Article number 21 regulates transnational practices. It states,

1. States, public and private institutions, and professionals associated with transnational activities should endeavor to ensure that any activity within the scope of this Declaration, undertaken, funded or otherwise pursued in whole or in part in different States, is consistent with the principles set out in this Declaration.
2. When research is undertaken or otherwise pursued in one or more States (the host State(s)) and funded by a source in another State, such research should be the object of an appropriate level of ethical review in the host State(s) and the State in which the funder is located. This review should be based on ethical and legal standards that are consistent with the principles set out in this Declaration.

---

[17] Andorno (2009b, p. 263).
[18] Andorno (2009b, p. 271).
[19] Bujo (1992, pp. 122–123).

3. Transnational health research should be responsive to the needs of host countries, and the importance of research contributing to the alleviation of urgent global health problems should be recognized.
4. When negotiating a research agreement, terms for collaboration and agreement on the benefits of research should be established with equal participation by those party to the negotiation.
5. States should take appropriate measures, both at the national and international levels, to combat bioterrorism and illicit traffic in organs, tissues, samples, genetic resources and genetic-related materials.

4.1.1.4   Promotion of the Declaration[20]

Being an all-encompassing universal norm, regulation and ideal; and being a theory, ideal and praxis, Ubuntu transcends national boundaries into the essence of humanity that all members of the species share. Exploitation is against Ubuntu whether it is between few members of the society or between national states.

Article number 22 empowers and encourages states to implement the principles of the Declaration. It as well underlines implementation of article number 19 which concerns creation and utilization of ethics committees. It states,

1. States should take all appropriate measures, whether of a legislative, administrative or other character, to give effect to the principles set out in this Declaration in accordance with international human rights law. Such measures should be supported by action in the spheres of education, training and public information.
2. States should encourage the establishment of independent, multidisciplinary and pluralist ethics committees, as set out in Article 19.[21]

This article's requirements on the states regarding implementation remained a duty and an obligation of each member of the traditional African society. Leadership would naturally eventually oversee harmony and concordance within their societies, but every member of the society would be responsible for oneself and for others in matters of morals and good conduct.

Article number 23 urges states to provide "Bioethics education, training and information." The Declaration explains,

1. In order to promote the principles set out in this Declaration and to achieve a better understanding of the ethical implications of scientific and technological developments, in particular for young people, States should endeavor to foster bioethics education and training at all levels as well as to encourage information and knowledge dissemination programmes about bioethics.

---

[20] Andorno (2009b, p. 283).

[21] Andorno (2009b, p. 293).

2. States should encourage the participation of international and regional intergovernmental organizations and international, regional and national non-governmental organizations in this endeavor.[22]

For African traditional society, each moment and each event is an occasion of learning. Learning is always based on life experience. Particular cases would be remembered for many years and passed on to subsequent generations as warning, regulation or instruction regarding right behavior or right course of action.

Article number 24 underlines International cooperation. It stipulates,

1. States should foster international dissemination of scientific information and encourage the free flow and sharing of scientific and technological knowledge.
2. Within the framework of international cooperation, States should promote cultural and scientific cooperation and enter into bilateral and multilateral agreements enabling developing countries to build up their capacity to participate in generating and sharing scientific knowledge, the related know-how and the benefits thereof.
3. States should respect and promote solidarity between and among States, as well as individuals, families, groups and communities, with special regard for those rendered vulnerable by disease or disability or other personal, societal or environmental conditions and those with the most limited resources.

The recommendations made by this article synchronize with Ubuntu philosophy. Personal and societal cooperation for the sake of common good belongs to the meaning of Ubuntu. However, unfortunately international exploitation is rampant right from the times of slave trade. Nowadays slave trade has changed its appearance into the often hidden underground international exploitation in form of prostitution which takes advantage of financial vulnerability of the victims, experimentation on human subject in poor countries and similar imperialistic unethical practices. In this case UNESCO and Ubuntu could not agree more.

Article number 25 is on "follow-up action by UNESCO." The article states,

1. UNESCO shall promote and disseminate the principles set out in this Declaration. In doing so, UNESCO should seek the help and assistance of the Intergovernmental Bioethics Committee (IGBC) and the International Bioethics Committee (IBC).
2. UNESCO shall reaffirm its commitment to dealing with bioethics and to promoting collaboration between IGBC and IBC.[23]

One of the handicaps of UNESCO is its lack of authority to actually implement the Declaration. UNESCO's Declaration on Bioethics and Human Rights still remains contingent on national states. As is evident in this number, it seeks collaboration of the Intergovernmental Bioethics Committee and the International Bioethics Committee. It is like a toothless dog that cannot bite. This situation is especially regret-

---

[22] Andorno (2009b, p. 303).
[23] Andorno (2009b, p. 317).

table when the world is confronted by tragedies such as the Syrian one in which a national administration can decide to gas its own people to death. The crisis in Syria reflects societal need for Ubuntu philosophy.

4.1.2   Articles Regulating Individual Human Treatment

The ultimate beneficiary of the stipulations and regulations of the UNESCO Declaration on Bioethics and Human Rights is both the human race as species and as an individual who will live in an environment worth of his dignity. Thus the declaration gives several directions on the treatment of individual human beings. Hence, the scope of the Declaration is not limited to universal norms or guidelines; it is also for and about individual good.

4.1.2.1   Self Determination

Article 18 deals with decision-making and the bioethical issues around it. It stipulates,

1. Professionalism, honesty, integrity and transparency in decision-making should be promoted, in particular declarations of all conflicts of interest and appropriate sharing of knowledge. Every endeavor should be made to use the best available scientific knowledge and methodology in addressing and periodically reviewing bioethical issues.
2. Persons and professionals concerned and society as a whole should be engaged in dialogue on a regular basis.
3. Opportunities for informed pluralistic public debate, seeking the expression of all relevant opinions, should be promoted.[24]

Although traditional medicine is not formalized Ubuntu philosophy favors informed decision-making. However, owing to Ubuntu worldview, decision-making and informed consent is not a private affair. Not only the patient would receive information that would help him make informed consent, the extended family or the community in which the patient belongs would also be involved in the process and participate in the decision-making.

Article 5 of the Declaration addresses respect for "Autonomy and individual responsibility." The autonomy of persons to make decisions, while taking responsibility for those decisions and respecting the autonomy of others, is to be respected. For persons who are not capable of exercising autonomy, special measures are to be taken to protect their rights and interests."[25] African concept of personal autonomy is necessarily relational. It has to be relational because, as Gyekye states, "The person is constituted, at least partly, by social relationships in which he necessarily

---

[24] Andorno (2009b, p. 255).
[25] Andorno (2009b, p. 111).

finds himself."[26] Although personal life is real, its reality is only meaningful in the context of relationality.

As far as Africans are concerned, the reality of the communal world takes precedence over the reality of the individual life histories, whatever these may be."[27] Hence, consent that excludes the inescapable network of relationships that form an extended family or community is simply unrealistic. It is from this perspective Osuji states, "consent rests on the consensus reached in consultation with the group rather than on that by the individual patient alone."[28] In sum, African autonomy is realistically relational. This inescapable existential relationality of human personhood is the distinguishing and the greatest contribution of Ubuntu philosophy to the world.

Article 6 is closely related with article 5. It is on consent. It states,

1. Any preventive, diagnostic and therapeutic medical intervention is only to be carried out with the prior, free and informed consent of the person concerned, based on adequate information. The consent should, where appropriate, be expressed and may be withdrawn by the person concerned at any time and for any reason without disadvantage or prejudice.
2. Scientific research should only be carried out with the prior, free, expressed and informed consent of the person concerned. The information should be adequate, provided in a comprehensible form and should include modalities for withdrawal of consent. Consent may be withdrawn by the person concerned at any time and for any reason without any disadvantage or prejudice. Exceptions to this principle should be made only in accordance with ethical and legal standards adopted by States, consistent with the principles and provisions set out in this Declaration, in particular in Article 27, and international human rights law.
3. In appropriate cases of research carried out on a group of persons or a community, additional agreement of the legal representatives of the group or community concerned may be sought. In no case should a collective community agreement or the consent of a community leader or other authority substitute for an individual's informed consent.[29] According to Ubuntu philosophy, information that is necessary for ethical decision-making is provided to the individual who belongs to the community.

It is provided to a father of children, to a child of somebody, an uncle or aunt of someone, to a mother of someone or to a niece or nephew of someone. There is no way this individual will be treated in isolation from this network of relationships. As stated before, to be is to relate and to belong. Failure to belong and to relate is tantamount to annihilation. Informed consent, therefore, is provided to a person who is necessarily in the context of belonging and relating. In other words, it is

---

[26] Gyekye (1997, p. 38).

[27] Menkiti (1984, p. 171, 180).

[28] Osuji (http://digital.library.duq.edu/cdm-etd/item_viewer.php?CISOROOT=/etd&CISOPTR=1 62271&CISOBOX=1&REC=2).

[29] Andorno (2009b, p. 123).

provided by and through the extended family of the individual or the community in which the individual belongs.

Ubuntu ethics, which defines selfhood, personhood and individuality in terms of otherness, implies that reality is in unison. Human genre is a unity composed of a plurality of individuality. Basically, all individuals within the plurality are equal in dignity; so equal that each can only define his existence in terms of an-other. Consequently, any action that reduces a human person to a kind of means for an end is immoral. A human being who uses another person as a means is by his very actions not human, since one becomes human through other humans. Hence in Bantu languages we have phrases such as: "*Hana Utu!*" Swahili phrase which is literally translated as "He lacks humanness." The phrase implies that a person is so lacking in morality (evidenced by his actions) that he is not human (since only human beings are moral beings in essence). In most African languages morality is synonymous with humanness.

4.1.2.2   Inability and Vulnerability

Article 7 has instruction on the treatment of "persons without the capacity to consent." The Declaration instructs that, "in accordance with domestic law, special protection is to be given to persons who do not have the capacity to consent:

a. Authorization for research and medical practice should be obtained in accordance with the best interest of the person concerned and in accordance with domestic law. However, the person concerned should be involved to the greatest extent possible in the decision-making process of consent, as well as that of withdrawing consent.
b. Research should only be carried out for his or her direct health benefit, subject to the authorization and the protective conditions prescribed by law, and if there is no research alternative of comparable effectiveness with research participants able to consent. Research which does not have potential direct health benefit should only be undertaken by way of exception, with the utmost restraint, exposing the person only to a minimal risk and minimal burden and if the research is expected to contribute to the health benefit of other persons in the same category, subject to the conditions prescribed by law and compatible with the protection of the individual's human rights. Refusal of such persons to take part in research should be respected.[30]

African traditional ethics would never allow using a person as a means for another person. Every person is substantially equal to every other person. The vulnerable enjoy protection of everybody else in the community. Article 8 aims at protecting the vulnerable. The article urges respect for vulnerability and integrity of the vulnerable persons. It states, "In applying and advancing scientific knowledge, medical practice and associated technologies, human vulnerability should be taken into

---

[30] Andorno (2009b, p. 137).

account. Individuals and groups of special vulnerability should be protected and the personal integrity of such individuals respected."[31]

Indigenous African communities have always given precedence to the, sick, bodily or mentally challenged and children. In many ethnicities failure to protect, enable and prioritize such special groups would call upon the healthy a wrath of God. It is always considered a blessing to care for those who cannot care for themselves. Implicitly, Ubuntu would never condone any kind of exploitation of the vulnerable.

Article 9 emphasizes Privacy and confidentiality. It states,

The privacy of the persons concerned and the confidentiality of their personal information should be respected. To the greatest extent possible, such information should not be used or disclosed for purposes other than those for which it was collected or consented to, consistent with international law, in particular international human rights law.[32]

Ipso facto that African life is to a very large extent a shared life, privacy and confidentiality is not as important as it is in modern Western medical ethics. MacIntyre very skillfully provides the rationale for this state of affairs. He states,

The story of my life is always embedded in the story of those communities from which I derive my identity. I am born with a past; and to try to cut myself off from that past, in the individualist mode, is to deform my present relationships. The possession of an historical identity and the possession of a social identity coincide. Notice that rebellion against my identity is always one possible mode of expressing it."[33] Consequently, absolute privacy and confidentiality that may exclude family or immediate community is not possible. Equally important is the precedence of community over individual personal life. Senghor describes this reality artistically when he posits, "Negro-African society puts more stress on the group than the individual, more on solidarity than on the activity and needs of the individual, more on the communion of persons than on their autonomy. Ours is a community society."[34] There is an individual life which is a tiny portion of the whole community life, and the two (individual and community life) are inseparable. Bujo notes that there is a unanimous consciousness of the primacy of community life over individual life. He asserts, "Every member of the community, whether it be family, clan or tribe, knows that he or she only lives by the life of the whole, and that God and the Founding Ancestor are sources of life."[35]

4.1.2.3   Individual Good Against Common Good Dilemma

Article number 3 underlines Human dignity and human rights. Instructs that

1. "Human dignity, human rights and fundamental freedoms are to be fully respected.

---

[31] Andorno (2009b, p. 155).

[32] Andorno (2009b, p. 165).

[33] MacIntyre (1984, p. 221).

[34] Senghor (1964, p. 49, 93–94).

[35] Bujo (1992, p. 124).

2. The interests and welfare of the individual should have priority over the sole interest of science or society."[36] Ubuntu respects personal human dignity, fundamental freedoms and human rights in within the matrix of the society in which the individual belongs. Ubuntu differs in perspective with regards to article number 3 of the UNESCO Declaration on Bioethics and Human Rights. While the Declaration emphasizes the precedence and priority of individual interests and welfare, Ubuntu quite realistically refuses to disentangle the individual from his socio-geographical and historical contexts.

According to the philosophy of Ubuntu human dignity and the rights that accompany it are respected in a context of *Thou-I* relationship. The perspective is simply represented in the maxim, "I am because we are; and since we are, therefore I am."[37] Basically Ubuntu underlines the often unrecognized role of relatedness and dependence of human individuality to other humans and the cosmos.[38] There is no conflict but mutual symbiotic affirmation between an individual and the community. This mindset is hardly understood in the West. Since the Declaration recommend respect for diversity and plurality this worldview must be recognized, understood and respected. Its foundation is represented in Bujo's statement, "Individuals live only thanks to the community."[39]

The worldview mentioned above is not only a theory among Africans. It is an epistemological, psychological and ontological reality. Hence Bujo states "Africans do not think in 'either/or' but rather in 'both/and' categories."[40] An individual is not against the community but with and for the community. Some critics have argued that this mentality is hazardous to individual's identity and self-determination. "Recent research has proven conclusively that the group does not at all dissolve the ethical identity of the individual,"[41] on the contrary, the group affirms and enhances the individual. The Ubuntu existential philosophy constantly underlines the undeniable role of otherness to selfhood. Implicitly, Ubuntu recognizes the significance of the bioethical principles of justice, beneficence, and non-maleficence, which tend to be wrongly preceded by that of autonomy.

In practice a patient would always be accompanied by some members of the extended family/community. This would always be the norm since one's life does not belong solely to him/her. "According to Gikuyu ways of thinking," for example, "nobody is an isolated individual. Or rather, his uniqueness is a secondary fact about him; first and foremost he is several people's relative and several people's contemporary."[42] Because of this constant awareness of belonging, "The personal

---

[36] Andorno (2009b, p. 91).

[37] Mbiti (1970, p. 41)

[38] Chuwa   (http://digital.library.duq.edu/cdm-etd/document.php?CISOROOT=/etd&CISOPTR=154279&REC=9).

[39] Bujo (2001, p. 3).

[40] Bujo (2001, p. 1).

[41] Bujo (2001, p. 6).

[42] Kenyatta (1965, p. 297).

pronoun 'I' was used very rarely in public assemblies. The spirit of collectivism was much ingrained in the mind of the people."[43] Thus, it is a common practice for a doctor to tell the diagnosis of a patient to the patient's family before telling the patient himself. Usually this is done to solicit community or family support of the patient in accepting and dealing with the reality of his health condition.

Article 4 of the Declaration is on the principles of beneficence and nonmalefi- cence. It states, "In applying and advancing scientific knowledge, medical practice and associated technologies, direct and indirect benefits to patients, research par- ticipants and other affected individuals should be maximized and any possible harm to such individuals should be minimized."[44] The philosophy of Ubuntu prioritizes the sick, the challenged and the vulnerable. It is the way one treats other people, especially those who are weaker than oneself that defines the individual's morality.

Even though Africans do not have most of the technology referred to in article 4 of the Declaration, they do have in place moral regulation as per how the sick and the vulnerable should be treated. Exploitation of the sick is an abomination within African traditional society. Nursing homes are a new phenomenon in Africa and people run away from them. People would like to surround their sick or old with love and care. Vulnerability calls for more attention and protection.

4.2 Justice

One of the major components of the UNESCO Universal Declaration on Bioethics and Human Rights concerns justice. John Rawls explores the concept of justice as a complex theory.[45] Cunneen relates restorative justice and reparations in establishing truth and resolving conflict between both victim and offender while reintegrating them in the society.[46] Hans Kelsen demonstrates the difficulty of defining absolute justice, especially because justice is subordinate to, and defined by social order. Justice is, in his perspective, relative.[47] In his work "Religion without God, Social Justice without Christian Charity, and Other Dimensions of the Culture of Wars," Cherry argues that all secular bioethics is empty if devoid of religious objectives.[48] He perceives ethics as a means to a religious end. However, the ethical debate on human rights respects the universal primacy of the human person within the param- eters of the principles of justice. This component is based on two major concepts. The first concept concerns dignity and freedom within the matrix of the principles of justice and solidarity. The second concept concerns equality of human beings as a fundamental premise and both a requirement and objective of ethical discourse.

---

[43] Kenyatta (1965, p. 188).

[44] Andorno (2009b, p. 99).

[45] Rawls (1999).

[46] Conneen (2008, p. 365).

[47] Kelsen (1996, pp. 183–206).

[48] Cherry (2009,) pp. 277–299).

4.2.1   Dignity and Freedom

The first concept of justice is that all human beings are naturally entitled to human dignity and fundamental freedoms. Denying them such entitlements violates their humanity.[49] Human dignity "has a key role in international bioethics" because all ethics is based on, and revolves around it.[50] The UNESCO Universal Declaration on Bioethics and Human Rights emphasizes that respect for human dignity and avoiding any abusive decision that would compromise human dignity for the sake of society "is of paramount importance." The declaration noted, however, that in many cultures and traditions, family and the community are more important. Thus, "the primacy of the human person finds its limits in the principles of justice and solidarity."[51] The declaration intentionally linked bioethics and global problems such as access to quality health care, nutrition, drinking water, poverty and illiteracy to emphasize the global primacy of human beings.

Since human dignity and freedom should be reciprocated between individuals and the community and should be honored by both individuals and the community, the declaration introduced a new principle called "Social Responsibility."[52] Some critics deny UNESCO authority to set such universal standards or to even discuss ethics.[53] Most individual ethicists also face such criticisms.[54] There is need, however, to have universal standards for the sake of the common good of humanity. De Castro, Sy and Chin Leong raised the issue of the 'global poor' as an issue of social and distributive justice.[55] Hessler and Buchanan state that due to inequality in national economies and policies, distribution of healthcare is problematic.[56] However, healthcare being a human right, such impediment is a mere excuse.[57]

After exploring and comparing healthcare systems in different national economies, Callahan and Wasunna discourage commercialized healthcare in the interest of human dignity.[58] Market forces of supply and demand do not necessarily recognize human dignity. Commercialized healthcare often aims at profit maximization at the expense of human dignity and freedom of choice. Ubuntu culture, though without formal written principles, fully recognizes, respects, and defends human dignity in practice. This work explores how Ubuntu assures human dignity and freedom within society as the matrix, which discerns and assures justice. Within Ubuntu, human life is invaluable. Everybody should do everything possible to pro-

---

[49] Andorno (2009b, pp. 91–98).

[50] Andorno (2007, p. 153).

[51] Andorno (2009b, pp. 33–44).

[52] Andorno (2009b, p. 33–34).

[53] Zwart (2007, p. iii).

[54] Zwart (2007, p. iii).

[55] de Castro et al. (2011, pp. 292–293).

[56] Hessler and Buchanan (2002, pp. 84–95).

[57] Rhodes et al. (2002, pp. 84–95).

[58] Callahan and Wasunna (2006, pp. 247–274).

tect and safeguard human life and dignity. This Ubuntu perspective is an inspiration to modern trends in healthcare.

4.2.1.1 Ethical Conflict Between Human Dignity and Commoditization of Healthcare

Since market economy operates on the basic principles of supply and demand, commoditization of healthcare tends to compromise human dignity. In commercialized medicine caregivers tend to specialize in the most marketable fields of medicine and patients who can afford to pay for better care or higher quality care receive better healthcare than those who cannot afford it. Treating the United States as a case study, in the last century medicine began to depend much more on sophisticated and specialized technology.[59] The advances in medical knowledge and efficiency of technology made technology an appealing option in medical care. Gradually, specialization became entrenched in the system as doctors focused on particular aspects of health such as radiology, neurology, allergy, cardiovascular surgery, oncology and other specialties. Such advances contributed to the shift to understand health care as a free-market commodity. Soon afterwards fee-for-service became the norm and included the opportunity to buy health insurance.[60]

Once medicine became a commodity to be purchased, "insurance became particularly important in the United States as health care costs rose to cover the expenses of medical technology, education, specialization, staffing, and facilities."[61] Athena du Pre articulates the situation as follows:

> The premise of insurance is to pool resources so that expenses are spread over a great number of people, saving any subscriber from overwhelming debt. The premise assumes that most people will not require more than they contribute and that enough people will subscribe to establish an adequate treasury.[62]

With generous reimbursement of medical costs by third parties, physicians were autonomous in clinical decision-making that impacted on the care of their patients and did not have to worry about the impact of the cost of medical procedures and treatment choices. However, by 1960 health care was becoming increasingly expensive. Health insurance rates rose beyond the reach of many Americans.[63] By 2003, 41 million Americans had no health insurance.[64] At this point the harsh reality of market forces of supply and demand sidelining human dignity became more apparent.

Fiscal scarcity and the rapidly changing health care market resulted in a shift of health care organizations from being solely physician dominated, "guild-like system that depended upon diagnosis and treatment of the patient as an individual," to

---

[59] du Pre (2000, p. 38).

[60] du Pre (2000, p. 39).

[61] du Pre (2000, p. 39).

[62] du Pre (2000, p. 39).

[63] du Pre (2000, p. 40).

[64] Physicians' Working Group for Single-Payer National health Insurance. (2003)

an industrialized model. The industrialized model relies on population-based statistical evidence and fiscal resource availability to organize and to provide health care predictability. This shift made health care a business. Those who could not purchase healthcare had to go without.[65] Census report indicates that the number of Americans without health insurance has been rising.[66]

The National Center for health Statistics reports that in 1984, approximately 30% of the population was without coverage. In 1993 that figure had risen to over 38% and by 1996 it had risen to nearly 40%.[67] In the year 1997, there were 40 million Americans without health insurance for the whole year.[68] This is over 16% of the entire population of the country. Currently approximately 47 million Americans have no health insurance. Among those who have insurance, there are many who have heavy health care burdens despite their being insured. Under insurance, a scenario whereby only some conditions are covered by insurers is common among marginalized portions of the society.[69]

Underinsured people spend a disproportionate amount of their income on health care. According to a recent study, 45 million Americans live in families that spend more than 10% of after-tax income on health care.[70] Three Institute of Medicine studies reported that the most important determinant of access to health care is adequate insurance coverage.[71] Even geographic areas with a robust safety-net care system fail to provide access to health services to the same extent as having health insurance.[72]

Part of the reason for the increasing cost of health insurance is the linkage of health insurance and employment. Due to increasing cost, some employers abandoned provision of health insurance all together.[73] Another trend is cost sharing between employers and employees, in which case employers would pay a given portion while employees would pay a portion by themselves. Often this scenario resulted in some employees opting out due to the rising cost of insurance and cost of living.[74] According to Marie Conn, lack of insurance among the economically marginalized portions of the American population creates a vicious cycle. The poorer the population, the less coverage, since less coverage means paying out of pocket and since the poor tend to take more risks with their lives and are open to more risky situations, the poor tend to get sick more frequently and in higher numbers. Being sick more often and in higher numbers than their wealthier counterparts and having to pay more and more out of pocket results in an ever-worsening vicious cycle,

---

[65] Boyle et al. (2000, p. 10).

[66] Kuttner (1999b, pp. 163–164).

[67] Centers for Disease Control and Prevention. (2012).

[68] Kuttner (1999b, pp. 163–168).

[69] Aday (1993, pp. 50–55).

[70] Schoen et al. (2005, pp. w5-289–302).

[71] Institute of Medicine June (2003).

[72] Institute of Medicine (2002).

[73] Freudenheim (1999, pp. 248–252).

[74] Kuttner (1999a, pp. 248–252).

which in return compromises human dignity even more.[75] The group of people who most need the coverage become the most likely to be denied coverage; the higher their need for insurance coverage, the less the possibility of receiving any.[76]

Those with limited coverage end up being denied coverage where they most need it since insurance providers are conditioned by market forces geared toward profit maximization. Most insurers tend to exclude some occupations, forms of industry, geographical areas, people with pre-existing conditions or those prone to some sort of illnesses.[77]

The wealthiest portion of the population gets the best insurance coverage in the world since they receive their coverage as a contract with third party insurance companies.[78] Since it is a private contractual right, however, its provision is contingent on employment in companies that can afford to provide such access.[79] The access is conditioned by continuing employment. Unfortunately such kind of access is on the decrease due to rising costs of health care and the cut back on financing of health insurance by employers.[80]

Due to the severity of market forces' control of healthcare, there has been a lot of abuse and neglect, which in turn would compromise human dignity. This situation led to creation of the Emergency Medical Treatment and Active Labor Act (EMLATA) by Congress in 1986. EMLATA is a limited legal right of "anti-dumping." Creation of EMLATA is a response to the dumping of so many uninsured sick persons, some in life threatening conditions. EMLATA's objective is to ascertain that uninsured patients will receive at least a minimum standard of emergency care regardless their ability to pay out of pocket.[81] Anti-dumping, however, neither addresses chronic conditions nor provides for continuity of care after the emergency treatment.

This situation indicates a major flaw in the system. It reveals a counterproductive situation in which the essence of the problem is not dealt with but the outcome of the problem. The problem is lack of healthcare coverage. Instead of proactively preventing the crisis, the system provides for a safety-net that only deals with the crisis when it happens. Such a scenario is generally inefficient and in the long run uneconomical.[82] There is an obvious issue of injustice in such a system. The following section explores possibilities of true justice.

4.2.1.2   Rawls' Perspective of Justice

The objective of Rawls theory of justice is to offer a fairer alternative to traditional concepts of utilitarianism and perfectionism as foundational theories of justice. His start-

---

[75] Conn (1997, pp. 899–1000).

[76] Whitted (1993, pp. 332–337).

[77] Bettistella and Kuder (1993, pp. 6–34).

[78] Sultz and Young (1997, pp. 286–287).

[79] Etheredge et al. (1996, pp. 93–104).

[80] Kuttner (1999a, pp. 248–252).

[81] Emergency Medical Treatment and Active Labor (EMLATA) (1998).

[82] Showalter (1999, Chap. 4).

ing point is an imaginary hypothetical starting position, which would legitimize social contracts. Rawls conceives justice as fairness. His pursuit of fairness led him to the development of his two famous principles of justice: the liberty principle and the difference principle. Rawls identifies the primary subject of justice as the basic structure of society, or more specifically the way in which major social institutions distribute fundamental rights and duties and determine the division of advantages from social cooperation.[83] His concept of justice is a provision of a standard, which improvises for the possibility of assessing the distributive aspects of the basic structure of the society.[84]

Rawls' original position is imaged as a hypothetical ideal in which no one knows his place in society, his class position or social status, his fortune in distribution of natural assets and abilities, his intelligence, strength or any other endowment. This state of affairs ascertains that the fundamental agreements reached in it are fair, since Rawls' meaning of justice is fairness. In the original position, all parties involved are equal. All have the same rights in the procedure for choosing principles; each can make proposals and submit reasons for their acceptance. This hypothetical condition along with the "Veil of Ignorance" define the starting point of the principles of justice as those which rational persons concerned to advance their interests would consent to as equals when none are known to be advantaged or disadvantaged by social and natural contingencies.[85]

Rawls' imagined ideal of the original position entails what he called "Veil of Ignorance," that is, a virtual committee of rational but not envious persons who would exhibit mutual disinterest in a situation of moderate scarcity as they consider the concept of rightness. Such concept has to be general in form, universal in application and publicly recognized. Rawls claims that rational people will unanimously adopt his principle of justice if their reasoning is based on general considerations, without knowing anything about their own personal situation. Such personal knowledge might tempt them to select principles of justice that gave them unfair advantage.[86]

Rawls identifies two principles that he believes would be chosen by all participants under the veil of ignorance in the original position. He further contends that the principles must be arranged in a serial order with the first principle prior to the second so that they do not permit exchanges between basic liberties and economic and social gains.[87] The two principles require equality in governing the assignment of rights and duties and regulating the distribution of social and economic advantage.[88] The first principle is that, each person is to have an equal right to the most extensive scheme of equal basic liberties compatible with a similar scheme of liberties for others.[89]

The second principle is that social and economic inequalities are to satisfy two conditions: (a) they are to be attached to positions and offices open to all under

---

[83] Rawls (1971, p. 7).

[84] Rawls (1971, p. 9).

[85] Rawls (1971, p. 17).

[86] Rawls (1971, pp. 130–135).

[87] Rawls (1971, p. 63).

[88] Rawls (1971, p. 61).

[89] Rawls (1971, p. 60).

conditions of fair equality of opportunity; and (b) they are to be to the greatest benefit of the least advantaged members of society (the difference principle).[90] Thus, although the distribution of wealth and income need not be equal, it must be to everyone's advantage, and positions of authority and offices of command must be accessible to all.[91]

There is striking similarity between Rawls' concept of justice as fairness and the basic idea of Ubuntu justice. Ubuntu does not condone dangerous inequality that may reduce a person from his essential equality with other persons on one hand, while on the other hand Ubuntu is not socialism in the sense that it does allow difference and entitlement in ownership. The permissible difference, however, is not only to the advantage of the privileged but, especially, to the advantage of the marginalized. Rawls' theory, as is Ubuntu perspective, entails a mechanism which safeguards human dignity and essential human equality while allowing some realistic entitlement and liberty. There is imbedded in the system a safety-net which prevents the gap between the richest and the poorest from enlarging disproportionally.

The rationale for regulating the economic gap between the richest and the poorest is well explained by Schrecker. He argues that "Most scarcities that underpin health disparities within and among countries are not natural; rather, they result from policy choices and the operation of social institutions." Schrecker argues for "denaturalizing scarcity as a strategy for enquiry to inform public-health ethics in an interconnected world." In his view, most scarcity is man-made.

It results from wrong policy in distribution of natural resources or products of human labor between human individuals and between populations or geographical regions. Thus "denaturalizing scarcity represents a valuable alternative to mainstream health ethics, directing our attention instead to why some settings are 'resource poor' and others are not."[92]

Rawls' theory of justice as fairness is inherently a theory of caring justice in the sense that it recognizes and safeguards human equality, dignity, basic rights and the principle of subsidiarity. The principle of subsidiarity appreciates every person's contributions while, at the same time, encourages participation and protection of those who cannot participate. Basically, Ubuntu worldview is similar to Rawls' theory of justice. UNESCO Declaration on Bioethics and Human Rights is in many ways in agreement with Rawls' theory of justice. Rawls' justice is not against ethical liberalism; rather it regulates liberalism so that it is not disproportional thus unethical.

4.2.1.3   Nagel on Rawls' Concept of Liberalism

Nagel notes that "Rawls interprets both the protection of pluralism and individual rights and the promotion of socioeconomic equality as expressions of a single value—that of equality in the relations between people through their common politi-

---

[90] Rawls (1971, p. 83).
[91] Rawls (1971, p. 61).
[92] Schrecker August (2008, p. 600).

cal and social institutions." The foundation of justice rests in the basic structure of society. The kernel of such structure is human equality. If the structure "deviates from this ideal of equality, we have societally imposed unfairness, hence the name 'justice as fairness.'" Thus the society is responsible for the structure that either supports fair treatment of all its members or supports unfair treatment of some of its members, which ultimately becomes unfair treatment of all members of the society. The society as a corporate person is not exempt. To underline this structural ethical reality Nagel states that "a society fails to treat some of its members as equals whether it restricts their freedom of expression or permits them to grow up in poverty."[93]

Nagel does not only approve Rawls' theory of justice, he states that it is "the fullest realization we have so far of this conception of the justice of a society taken as a whole whereby all institutions that form part of the basic structure of society have to be assessed by a common standard."[94] Credibility of Rawls' theory of justice as fairness consists of the fact that it starts from scratch and at a point of imaginable ideal but also real and factual equality which should not be overlooked, even in the sophisticated and complicated structures of modern societies.

The theory then protects the essential common human values that all human beings share. It protects and defines human freedom in relation to fairness based on human inviolability. Nagel writes "The protection of certain mutual relations among free and equal persons, giving each of them a kind of inviolability, is a condition of a just society that cannot, in Rawls' view, be explained by its tendency to promote the general welfare. It is a basic, underived requirement."[95] The kernel of Rawls' theory therefore is equal human dignity which must be given its due fairness wherever humans are located geographically, socially and economically.

To be ethically justifiable the equality of human dignity which calls for its share of fair treatment should not overlook, undermine or suppress diversity, plurality and liberty. The first of Rawls' principles is thus one of irreducible and undeniable equality while the second principle is one that protects ethically reasonable and essential inequality. Nagel relates that

> Rawls' difference principle is based on the intuitively appealing moral judgment that all inequalities in life prospects dealt out to people by the basic structure of society and for which they are not responsible are prima facie unfair; these inequalities can only be justified if the institutions that make up that structure are most effective in achieving an egalitarian purpose—that of making the worst-off group in the society as well off as possible.[96]

In praxis an affluent society bears ethical responsibility of ascertaining that the disadvantaged children born to a poor family gets all basic needs and the education they need to have a fair chance to self-actualize and be free to excel just as children of the wealthy members of the society. In other words, if the poor keep getting poorer and keep being deprived of chances to get out of their poverty even if they

---

93 Nagel (2003, p. 65).

94 Nagel (2003, p. 63).

95 Nagel (2003, p. 65).

96 Nagel (2003, p. 71).

would want to; if they are not enabled by their wealthier counterparts because the structure does not support it, the whole socio-economic structure is unethical.[97]

People ought not be systematically rewarded or penalized "on the basis of their draw in the natural or genetic lottery." The only way to justify difference is to ensure that "the system works to the maximum benefit of the worst off" because, as Nagel articulates, "People do not deserve their place in the natural lottery any more than they deserve their birthplace in the class structure, and they therefore do not automatically deserve what 'naturally' flows from either of those differences."[98]

Rawls' justice neither disregards nor ignores human plurality. Interpreting Rawls, Nagel writes "that pluralism and toleration with regard to ultimate ends are conditions of mutual respect between citizens that our sense of justice should lead us to value intrinsically and not instrumentally." However, the "Veil of Ignorance" is crucial since it protects the basic commonality and equality of human nature without undermining accidental differences. Interpreting Rawls Nagel writes, the "feature of the veil of ignorance, like not knowing one's race or class background, is required because Rawls holds that equal treatment by the social and political systems of those with different comprehensive values is an important form of fairness."[99]

Plurality is to the advantage of the society. According to Rawls, "A wide range of views, forming the plurality typical of a free society, are reasonable and can support the common institutional framework." Rawls calls this ethically justified plurality "an 'overlapping consensus'." Which means the uniqueness and the simultaneous compatibility of each of the "comprehensive views with a free-standing political conception that will permit them all to coexist."[100]

Rawls' theory of justice, therefore, cannot be ignored by those who are concerned with social justice. The UNESCO Declaration on Bioethics and Human Rights, either directly or indirectly is inspired by, or has a lot in common with Rawls' theory of justice as fairness. Trying to justify the objectivity of Rawls' theory of justice, Nagel writes "Rawls has not only expressed a distinctive position but provided a framework for identifying the morally crucial differences among a whole range of views on the main questions of social justice."[101] Needless to say, Rawls' theory of justice has a lot in common with Ubuntu perspective of justice. The imbedded socio-autonomous recognition of human essential equality to be protected; the importance of recognition and use of difference and plurality; especially how difference should be to the advantage of the most disadvantaged (by genetic pool or other factors) almost equate Rawls' theory of justice with the indigenous Ubuntu perspective of justice.

---

[97] Nagel (2003, p. 69).

[98] Nagel (2003, p. 72).

[99] Nagel (2003, p. 73).

[100] Nagel (2003, p. 84).

[101] Nagel (2003, p. 72).

4.2.2   Equality

The second concept of justice is the acknowledgement of universal human equality and equity, which is fundamental in ethics discourse on all that impacts humans regardless of their uniqueness and difference.[102] D'Empaire notes that the "principles of equity, justice and equality are basic in ethics and they have to be considered as part of any ethical system."[103] This statement is consistent with article 10 of the UNESCO Universal Declaration on Bioethics and Human rights which states that, "The fundamental equality of all human beings in dignity and rights is to be respected so that they are treated justly and equitably."[104] The Declaration recognizes and emphasizes human equality which should lead to treating each human being with equity and justice.

However, basing their argument on the draft of the declaration, Rawlinson and Donchin argue that the formulation of the universal principles of the UNESCO Declaration on Bioethics and Human Rights relies solely on shared ethical values while ignoring the differences which occur as a result of different cultures and fixed structural economic differences. They contend that the UNESCO Declaration on Bioethics and Human Rights is too abstract to be applicable.[105]

Dan Beauchamp makes a case against the commercialization and commoditization of healthcare.[106] From his perspective, commoditization of healthcare works against human equality. However, the challenge of translating theoretical understanding into real, practical life situations confronts all of human society. Presently, many populations are denied basic human rights throughout the globe.[107] Some marginalized people have been used as a means to an end by other humans. McDonald and Preto address this ethical problem in the area of global health research as conflict of interest. Daniels explores the global crisis of inequality in healthcare in depth.[108] Inequality in healthcare is an issue of justice which results in the denial of human equality to the victims.[109] Ubuntu worldview helps review the importance of assuring basic human equality for human common good.

Inequality in healthcare distribution remains a global problem even if healthcare is considered a human right that reflects respect for human dignity. Although some governments have ways to regulate healthcare distribution in order to ascertain the decent minimum for all, the still problem of unequal distribution remains. Ubuntu recognizes the equal dignity of humans in a rather practical way. Every human being has something to offer to every other human being, even if it is provision of an opportunity to help. One's very personhood is based on the recognition of other persons as equals to oneself and as participants in the formation of one's personhood.

---

[102] d'Empire (2009, pp. 173–185).

[103] d'Empaire (2009, pp. 175–176).

[104] d'Empaire (2009, p. 173).

[105] Rawlison and Donchin (2005, pp. 1471–8731).

[106] Beauchamp (1988, pp. 31–68).

[107] d'Empaire (2009, pp. 180–182).

[108] Daniels (2008, p. 333).

[109] McDonald and Preto (2011, pp. 327–329).

In Ubuntu culture it is the responsibility of everyone to ascertain the provision of decent minimum of care for all. Healthcare in Ubuntu reflects reverence for life as a matter of religion, morality and essence of humanity.

4.2.2.1 Castro, Sy and Leong on the Global Need to Address Dehumanizing Poverty

According to Castro, Sy and Leong extreme poverty and destitution among indigenous peoples is a global responsibility.[110] Its mere presence indicates unjust global socio-economic distributive structures. Morally, rich countries, individuals and corporations cannot exempt themselves from the plight of the global poor. Sy and Leong contend that "A corporation's responsibility to address the health needs of the poor extends beyond the country in which it directly operates. It has to be concerned with the global implications of its operations and not merely be preoccupied with the limited impact at the national or community level."[111] Sy and Leong's approach is cosmopolitan in the sense that in their view, countries, corporations and individuals belong to a global community. Cosmopolitanism contends that "distributive justice applies globally, not simply nationally or locally; therefore, there are moral obligations to address the plight of the poor of the world as a whole."[112]

The mere existence of abject poverty facilitates a moral slippery slope whereby the poor are forced by their poverty to become poorer to the point of being exploited in their very humanity. Having no way out, the poor populations may easily be forced to become a means to an end for the rich. Organ transplantation trade is a good example. Sy and Leong observe that "Massive poverty in developing country communities has provided the backdrop for debates regarding compensation for organ donors. In some communities, organ selling has reached wholesale proportions, making organ trading a literal reality." This situation demonstrates how poverty may set humans into a slippery slope of moral degradation whereby human dignity is compromised. In this case the poor are literally used as a means to the ends of the rich.

"Patients from affluent foreign countries have exploited the opportunities that are ably facilitated by clandestine brokers, thus setting in motion a practice that has straddled the boundary between transplant tourism and organ trafficking."[113] There is structural injustice when humans are forced to become a mere means; or where the situation is that of struggle for survival and survival of the fittest or strongest since such situations drain the essence of humanity by compromising its dignity. This degradation of humanity does not merely apply to the exploited poor. It applies to the entire human community.

Organ transplantation trade may lead to a situation morally similar to slave trade since people who would otherwise not give up their organs are forced by their pov-

---

[110] de Castro et al. (2011, pp. 291–292).
[111] de Castro et al. (2011, p. 292).
[112] de Castro et al. (2011, p. 292).
[113] de Castro et al. (2011, p. 297).

erty to do so against their will for sake of survival. According to Sy and Leong "organs such as kidneys and livers must be regarded as sacrosanct and outside the realm of commerce. On this basis many hold that organ donation must always be motivated only by altruism," especially because of irreducible human dignity. "Monetary considerations demean human donors and transform their bodies into commodities that can be reduced to a monetary or material equivalent."

It is because of the urgency to avoid the inevitable compromise of human dignity through human organ trade that "the Declaration of Istanbul on Transplant Tourism and Organ Trafficking rejects 'transplant commercialism' as 'a practice in which an organ is treated as a commodity.'"[114] Any form of directly or indirectly forced commercialism on human tissue or organ is unethical. If poverty makes people sale their own members, poverty is a structural moral evil that human community has to eradicate.

No government should prohibit its marginalized populations from engaging in illegal human organ trade if the government cannot provide for their basic need to survive. This means the problem of organ transplantation trade is much more complicated than it may look. It is a structural problem. Wherever it is happening, the immediate society and ultimately the global human society is responsible and culpable. Sy and Leong state that "Society that deliberately and systematically neglects the basic needs of the poor is being indifferent to the plight of this population and cannot be justified in prohibiting the means the poor have to address the problems themselves."[115] Consequently, prohibiting organ transplantation trade should be preceded by addressing the root cause of such dehumanizing trade, which is poverty. Ubuntu maxim that human beings are human because of other humans, or put briefly, "I am because you are" means that no one is free from the plight of other humans. Claiming such freedom from others would mean claiming inhumanity.

4.2.2.2 Beauchamp on the Ethical Need for Basic Human Equality in Medicine

The foundation of Beauchamp's argument is human equality within the state. His main premise is "common membership in a republic of equals." It is human equality and common membership of citizens in a republic that is the foundation of healthcare distributive justice. "Illness is the relevant reason for distributing medical care and health protections." The daunting ethical task is discernment and determination of the most ethical "pattern of organization of equality we ought to employ to make the equal distribution of medical care effective."[116]

According to Beauchamp the distinguishing feature of a central government is its duty to protect public health based on citizens' equality. In other words, it is unjust for a republic's government to fail to safeguard both equality and health of its citizens.[117] Beauchamp emphasizes on the objective of a republic as attempting "to

---

[114] de Castro et al. (2011, p. 297).
[115] de Castro et al. (2011, p. 298).
[116] Beauchamp (1998, pp. 2–3).
[117] Beauchamp (1998, pp. 11–12).

foster a sense of common membership and community." In his view "community like friendship, family, kinship, fraternity, and patriotism, refers to shared sentiments and attachments that bind people or groups to one another.

A republic, with its stress on virtue and a shared common life, is a species of political community."[118] Consequently, the ethical government's goal is to create, foster, and protect a community of equals. In order to achieve common good and promote harmony and equality the republic has a duty to limit individual liberty.[119] Thus defining and limiting individual liberty belongs to the kernel of justice. Beauchamp states that "Justice, in my account, is based not only on considerations of what each citizen needs but also on considerations for what everyone needs together."[120]

In as much as human equality is undeniable, widening gap between the rich and the poor that tends to indicate essential human inequality is obviously unethical, unjustifiable and intolerable. Such dehumanizing gap is unethical specifically because it is unreal and untrue. Beauchamp explains this as follows: "The very obviousness of a common and shared equality is the political glue for equality and justice in health, making it more difficult to island the poor, commercialize medicine, or allow an uncontrolled and expensive medical technology to erode further the society's commitment to equality in health."[121]

The greatest single threat to human essential equality in healthcare is the ongoing commercialization of healthcare. Commercialization of healthcare is commoditization of healthcare. Commoditization of healthcare gives market forces of supply and demand precedence over human dignity. Beauchamp explains this fact in a more practical way when he states that "As medicine moves deeper into the stronghold of the market, justice for the poor and the vulnerable will be increasingly unstable and the politics of a democratic majority moving to a common health care system may be permanently undermined."

In other words, the healthcare system is becoming unethical because it is being influenced and motivated by wrong objectives: the market. Beauchamp refers to this ethically dangerous phenomenon when he states that "the health care system, far from serving as a symbol of shared equality, is rapidly becoming a symbol of inequality."[122] One of the most obvious examples is the tendency to tend to deny coverage to those most in need due to profit maximization motive that has infiltrated healthcare. Beauchamp observes that "it is the ordinary and rational insurance practice to eliminate wherever possible from coverage, the highest utilizers of care, that is, ironically, those who most need care."[123]

Beauchamp laments that Americans resist health reform because, "we wish to provide a welfare state without the inconvenience of limiting the market." Unfortunately it is not possible to have both scenarios. "We will have to decide soon, perhaps for all time, whether we want a just health care system or market institutions that spread

---

[118] Beauchamp (1998, p. 15).
[119] Beauchamp (1998, p. 22).
[120] Beauchamp (1998, p. 40).
[121] Beauchamp (1998, p. 40).
[122] Beauchamp (1998, p. 47).
[123] Beauchamp (1998, p. 51).

to every corner of American life. Our choice will have profound consequences for healthcare, for equality, and for the American republic."[124] Opting to subject humans under the mercy of market forces is obviously unjust to human common dignity and equality. Human equality, however, ought not to undermine individual pursuit of individual good. Beauchamp explains how best to pursue individual interest ethically. His explanation is concomitant with Ubuntu perspective. He states that "In republican equality we promote our own good and our shared common good within the same democratic scheme."[125] Since individuals humans are inseparable from society because of their social nature and neediness for society, individual pursuit of fulfillment and happiness cannot be separate from societal objectives for the common good.

4.2.2.3   Daniels on Ethics of Ignorance and International Harm in Healthcare

According to Daniels, there is an obvious colossal injustice within the global healthcare system. This global injustice within healthcare though global responsibility is ignored by individual persons, corporations and states. To explicate global inequality and injustice in health care Daniels uses the following data:

> Life expectancy in Swaziland is half that in Japan. A child unfortunate enough to be born in Angola has seventy-three times as great a chance of dying before age five as a child born in Norway. A mother giving birth in southern sub-Saharan Africa has 100 times as great a chance of dying in labor as one birthing in an industrialized country.
> For every mile one travels outward toward the Maryland suburbs from downtown Washington, D.C., on its underground rail subsystem, life expectancy rises by a year—reflecting the race and class inequalities in American health.[126]

Health inequality between social groups according to Daniels results from "an unjust distribution of socially controllable factors that affect population health and distribution." Health inequalities follow different but often times common patterns. Often health inequalities are "by race and ethnicity, by class and caste, and by gender—in many countries, both developed and developing."[127]

Most of the harm to the poor peoples of the world results from ignorance of the rest of world's population about its obligations to the poor, ignorance of human rights and the need to respect them, and insensitivity to the plight of the poor. Daniels argues that "health of citizens of a specific nation is a responsibility of that specific nation. However, there are international breaches of human rights in form of omission or ignorance from wealthy nations to poor nations." There are other oppressive or exploitative practices which are unjust to the poor and which marginalize them even more but often go on unnoticed. Some of those injustices are: hazardous waste

---

[124] Beauchamp (1998, p. 67).
[125] Beauchamp (1998, p. 132).
[126] Daniels (2008, p. 333).
[127] Daniels (2008, p. 334).

disposal from industrialized countries in poor developing countries, international policies that intentionally or unintentionally harm poor countries, and brain drain.[128]

Brain drain by the global affluent countries from poor countries is worth attention since it is not only global ethical challenge, it is growing rapidly. "Rich countries have harmed health in poorer ones by solving their own labor shortages of trained health care personnel by actively and passively attracting immigrants from poorer countries." For individual survival or gain, the poor struggle to leave their poor countries to find a better life in the developed countries. Unfortunately, those who can even afford to think of that migration are the well trained ones. Their leaving their own countries harms the countries which have spent their little fortune to educate them. Such poor countries are doubly harmed as they are forced by harsh realities of market forces to let go of what they need the most. On the other hand countries which already have many health professionals benefit by gaining even more supply. "In developed countries such as New Zealand, the United Kingdom, the United States, Australia and Canada, 23–24 % of physicians are foreign-trained. In 2002, the National Health Service in the United Kingdom reported that 30,000 nurses, some 8.4 % of all nurses, were foreign trained."[129]

International brain drain leaves the donor countries in a humanly unethical shape. The situation that results after brain drain in developing countries is dire. Over 60 % of the doctors trained in Ghana in the 1980s emigrated oversees. In Ghana, 47 % of physicians' posts and 57 % of registered nursing positions were unfilled. Some 7,000 expatriate South African nurses work in developed countries, while there are 32,000 public health nursing vacancies in South Africa. Whereas there are 188 physicians per 100,000 population in the United States, there are only 1 or 2 per 100,000 in large parts of Africa.[130]

Even though some of the brain drain is not intentional, the harm is obvious as seen in the above figures provided by Daniels. Although the intent to harm is rarely present, the benefit if often times intended. Some developed countries even give incentives to attract professionals into their countries, regardless the harm done to the donating countries. The severity of the harm done to the economies and the people of donor countries can hardly be accurately measured:

> In any case, great care must be taken to describe the baseline in measuring harm. Such a complex story about motivations, intentions, and effect might seem to weaken the straight-forward appeal of the minimalist strategy, but the complexity does not undermine the view that we have obligations of justice to avoid harming health.[131]

Internationality and grandiosity of the brain or talent drain should not conceal its essential injustice. There is need to address this growing international problem.

Permanent or long term solution of the problem of brain or talent drain lies in recognizing human equality and addressing the core causal factors. Daniels explains the need to "move beyond minimalist strategy that justifies only avoiding and cor-

---

[128] Daniels (2008, p. 338).
[129] Daniels (2008, p. 338).
[130] Daniels (2008, p. 338).
[131] Daniels (2008, p. 340).

recting harms. How far we go toward robust egalitarian considerations is a matter to be worked out." However, egalitarian perspective is crucial if at all solution is to be found and maintained.

There is need to develop national and international institutional structures, based on human equality to discourage unethical brain and talents drain.[132] Just health cannot be an exclusive pursuit of an individual person or nation. As Daniels puts it, it is individual, societal, national and international pursuit. There is an essential unity of human genre which cannot be denied.[133] Ubuntu warns that no humanity is possible independent of human relationships. This inspiration is not limited to unique individuals; it applies to the entire global human community. Reducing any human individual or nation to a means for another individual or nation harms the essence of human nature and its dignity.

4.3   Diversity

The second major component of the UNESCO Universal Declaration on Bioethics and Human Rights is based on diversity. The debate on ethical responsibility must respect cultural and racial diversity within a global context. Respect for diversity has two important concepts. The first concept concerns cultural pluralism within the limits of human rights. The second concept concerns nondiscrimination based on essential human equality.

4.3.1   Cultural Pluralism

UNESCO advocates for respect for cultural pluralism based on, not at the expense of, human dignity.[134] Article 12 of the UNESCO Universal Declaration on Bioethics and Human rights clearly recognizes the importance of cultural diversity. However, the article indicates that cultural values are secondary to human rights. Universal human rights "guarantee the particular expression of individual cultures."[135] Human rights should, on the other hand, limit and provide for boundaries with respects to cultural pluralism. The UNESCO Declaration on Bioethics and Human Rights is founded on a basic assumption of human solidarity.

Gunson describes basic solidarity as "the willingness to take the perspective of others seriously, which in turn entails acting in ways that support the causes that are worthy of allegiance."[136] Responding to criticism that UNESCO Universal Declara-

---

[132] Daniels (2008, p. 354).

[133] Daniels (2008, p. 334).

[134] Revel (2009).

[135] Gunson (2009, p. 256).

[136] Gunson (2009, p. 256).

tion on Bioethics and Human Rights is a form of cultural imperialism, Andorno argues that the declaration actually works against cultural imperialism.[137] It provides "a legal standard of minimum protection necessary for human dignity." There is a general trend to global cultural integration which begs for such a universal standard.

Chin and Starosta explore in depth the relationship between modern technology, globalization, economy, wide-spread population migrations, cultural integration, development of inevitable multi-culturalism in the context of global culture and the role of effective communication.[138] In itself, globalization necessitates better and more effective cooperation between nations and peoples in meeting the legal standard of care for all people.[139] The role of the principles of bioethics is crucial in discerning and regulating conflict between freedom of cultural practices and respect for basic human rights regardless of specific national laws and boundaries.[140]

The culture of Ubuntu flourishes in diversity and pluralism. The ability to go beyond oneself to embrace others is an ethical ideal of conduct. Since beings become persons because of others and because relationships facilitate recognition and respect for personhood in each other, otherness and its plurality is richness. This component of Ubuntu was explored detail in Chap. 2. Ubuntu is thus enlightened by the UNESCO Declaration on Bioethics and Human Rights while it simultaneously provides sample praxis of the relevance of UNESCO's ethical directives.

4.3.1.1   ten Have on Solution of Moral Problems by Negotiation

ten Have notes that bioethics is becoming increasingly international even though many countries in the developing world do not have "adequate infrastructure to deal with bioethical issues" such as "expertise, ethics committees, ethics teaching programs, and ethics-related regulations and legislation." One of the reasons that ten Have points out for this awakening internationality of bioethics is the fear of the developing world to be "excluded from the benefits of biomedical progress."

ten Have cautions against the possibility of "double, or at least different, moral standards being applied in different regions of the world."[141] ten Have's warning is important, especially because of the cultural pluralism. Even though pluralism of perspectives is enrichment to global bioethics, there are ethical constants that must remain always universally objective regardless cultural perspectives. Double standard in bioethics relativizes it, thus compromising its validity.

Given the globalization of bioethics in the plurality of world cultures, there is need for negotiation. Basing his main reference on Beauchamp and Childress, ten Have critically analyses the main trends which should be considered in global bioethical negotiation. He explores foundationalism, antifoundationalism, common morality,

---

[137] Andorno (2009a).

[138] Chen and Starosta (2008).

[139] Taylor (2002, pp. 975–976).

[140] Revel (2009, p. 200).

[141] ten Have (2011, p. 20).

principles and Fallibilism. Since each school has both proponents and opponents, there is need for negotiation. Proponents of bioethical foundationalism hold that some bioethical principles "can be based on noninferentially justified beliefs." Such principles can thus "be rationally defended and they apply to all human beings." Proponents of foundationalism hold that "bioethical judgments can only be justified on the basis of an ethical theory that is rational and universal at the same time."[142] Foundationalism is crucial not only because of its belief in universal principles but also because of its unifying perspective which appeals to rationality and human nature.

The opposite of foundationalism, antifoundationalism, holds that "there are no ethical principles that are certain and universally valid, so that all moral judgment can be firmly grounded on them."[143] Since this view tends towards concreteness and uniqueness of moral situations, it holds that bioethics should be less universalistic, less generalizing and more "appreciative of the actual experiences of practitioners and more attentive to the context in which physicians, nurses, patients and others experience their moral lives." This perspective defends the unique, historical, cultural, abstract, relational and rational nature of bioethical encounters. Antifoundationalism holds that "persons are always persons-in-relation, are always members of communities, are immersed in a tradition, and are participants in a particular culture."[144] Antifoundationalism is concomitant to most ethics of care because of its emphasis on concreteness and uniqueness as opposed to universality and objectivity.

Common morality view tends to defend the innate nature of morality. ten Have refers and elaborates this tendency when he states "Before acting morally we must already know, at least to some extent, what is morally desirable or right. Otherwise, we would not recognize what is applicable in moral sense." Hence, human beings are naturally moral beings and that "moral normativity is pre-given and common to all human beings."

This position tends to bring together foundationalism and antifoundationalism since it recognizes both universality and historicity of moral precepts. Even though humans have innate knowledge of right and wrong, or good and bad "what we recognize in our experience is typically unclear and in need of further elucidation and interpretation."[145] Unlike foundationalist perspective, common morality perspective recognizes both universality of moral principles and the role of history and context.[146] "Cultures differ but this does not imply that common standards and universal principles do not exist."[147]

Principles and Fallibilism holds that "ethical principles do not have a stable and immutable foundation, but they need justification. Moral principles are justified if they contribute to the objectives of morality, such as human flourishing." Thus

---

[142] ten Have (2011, p. 23).
[143] ten Have (2011, p. 24).
[144] ten Have (2011, pp. 24–25).
[145] ten Have (2011, p. 26).
[146] ten Have (2011, p. 27).
[147] ten Have (2011, p. 27).

moral principles are rightly a means to an end because, in themselves, moral principles are useless. Morality should be at the service of human flourishing.

However, principles and Fallibilism tend to make moral principles conventional and fluid. One of the advantages of this position is its openness and welcoming stance to cultural contribution into justification of moral principles for the sake of human flourishing.[148] This view of morality encourages dialogue and development of moral theories since it constantly engages them by its demand of justification. However, it tends to compromise universality of moral principles.

ten Have observes that there is tendency toward more negotiation with regards to ethical principles. He writes, "Deliberative democratic processes are replacing the search for universal solutions that can be applied to all human beings. However, the significance of deliberation does not restrict the universality of ethical principles. Solutions to moral problems are no longer found and based on fundamental theories but are now negotiated."[149] In order to ethically respond to the demands of globalization of bioethics negotiation with indigenous and different cultures is crucial. ten Have writes that "UNESCO strives to respond in particular to the needs of developing countries, indigenous communities and vulnerable groups of persons. The declaration reminds the international community of its duty of solidarity toward all countries."[150] This desire of UNESCO to respond to the particular needs of developing countries requires common mutual understanding which in turn requires effective cultural dialogue, negotiation and understanding. The requirement of mutual recognition and engagement belongs to the core of Ubuntu world view.

4.3.1.2 Walzer on Pluralism and Distributive Justice

Pluralism has a lot in common with distributive justice. In fact acceptance of pluralism is not possible without, at the same time, an acceptance of validity of distributive justice. Walzer validates this perspective when he argues that "the idea of distributive justice has as much to do with being and doing as with having, as much to do with production as with consumption, as much to do with identity and status as with land, capital or personal possessions."[151] In other words, Walzer argues for the centrality of the importance of distributive justice in social ethics.

Walzer sums up this perspective when he states that "distribution is what social justice is about."[152] Nothing escapes the realm of distributive justice. Even the community itself is subject to distributive justice. Walzer argues, "The community itself is a good—conceivably the most important good—that gets distributed. But it is a

---

[148] ten Have (2011, p. 28).
[149] ten Have (2011, p. 28).
[150] ten Have (2011, p. 20).
[151] Walzer (1983, p. 3).
[152] Walzer (1983, p. 11).

good that can only be distributed by taking people in, where all the senses of that latter phrase are relevant: they must be physically admitted and politically received."

Thus, there is a different kind of distribution when applied to the community because humans become members of the community, thus being encompassed by it and becoming part of it, "hence membership cannot be handed out by some external agency; its value depends upon an internal decision."[153] Nevertheless human community is an ethical good that is unique for its grandiosity and whose distribution is by membership into it. In fact the community as a good is a prerequisite and a condition for all other forms of distribution.

According to Walzer need is the most basic reason for distributive sphere. "Need generates a particular distributive sphere, within which it is itself the appropriate distributive principle." Fairness requires that basic needs are mate with fair distribution relative to availability of the needed good. Distributive justice does not necessarily require uniformity. Just as plurality is complicated so is distribution, and even more is distributive justice. Distributive justice is complicated by scarcity of basic needs by different people.

Walzer refers to this fact when he speaks of "needed goods distributed to needy people in a proportion to their neediness are obviously not dominated by any other goods." Distributive justice should always be based on human equality, need, and plurality. It should be, as Walzer writes "different goods to different companies of men and women for different reasons and in accordance with different procedures." In Walzer's words, this statement contains the basic objective of the principle of distributive justice. He states, "To get all this right, or to get it roughly right, is to map out the entire social world."[154]

Most social conflict arises from unfair or ineffective distribution. Walzer argues that social justice is "intermittent, or it is endemic; at some point, counterclaims are put forward." There are three major kinds of counter claims worth noting:

1. The claim that the dominant good, whatever it is, should be redistributed so that it can be equally or at least more widely shared: this amounts to saying that monopoly is unjust.
2. The claim that the way should be opened for the autonomous distribution of all social goods: this amounts to saying that dominance is unjust.
3. The claim that some new good, monopolized by some new group, should replace the currently dominant good: this amounts to saying that the existing pattern of dominance and monopoly is unjust.[155]

Due to individual human and cultural uniqueness human society is inevitably pluralistic. It is pluralism that calls for just distribution. One of the major challenges facing UNESCO is to design an international model of distribution that will be just across nations. This ideal may not be easily achievable due to the different individual national identities and needs, but the closer the international community is to this

[153] Walzer (1983, p. 29).

[154] Walzer (1983, p. 26).

[155] Walzer (1983, p. 13).

ideal objective, the more just the world would be. The farther any particular nation or community of world nations deviates from the ideal of fair distribution, the more conflicts will multiply and the more human dignity is compromised.

Ubuntu aims at this ideal by linking morality with human ability to empathize and responsibly address the need of another human being, thus effecting distribution in a relational and engaging way. The society expects every person to actively participate. This kind of responsible participation is considered moral maturity. Actually, personhood is based on this sort of ethical maturity. In a very spontaneous physically coercive way Ubuntu ascertains fair distribution without encouraging uniformity or discouraging personal initiative and excellence.

4.3.1.3 Amstutz on the Ethics of Global Society and Governance

Amstutz raises one of the most disabling aspects of 'international community.' There is a definition problem with regards to referring to the nations of the world as an 'international community,' because the bonds that are necessary between nations are too weak and sometimes inexistent or hostile to deserve the word 'community.' According to Amstutz "The international community remains a society of states in which ultimate decision-making authority rests in member states, not intergovernmental organizations or non-governmental organizations."

Strictly speaking, therefore, there is no international community of nations as such. "Some officials use the phrase 'international community' to refer to actions by the United nations and other intergovernmental organizations, the level of solidarity among states and the degree of communal bonds among nations remain weak ... global society is held together by feeble institutions and slender affinities."[156]

Due to the lack of real communal solidarity and a central government there is really no real authority that oversees issues of justice between or within government with ability to intervene. United Nations and its agencies do not have such authority. They can only play an advocacy role. Amstutz points out one of the world's institutional limitations as the ever widening economic gap between rich (North) and poor (South) nations. The second example is the obvious world's failure to maintain global peace. "When major disputes arise between states, it is states themselves who must resolve conflicts, either directly or through intermediaries." Another example is the "inadequate protection of human rights." Yet another piece of evidence is the protection of the environment. Amstutz notes that "although numerous multilateral efforts have been undertaken to protect the environment, the decentralized character of global society impairs effective collective action."[157]

According to Amstutz several factors impede institutionalization of global governance. "One impediment is the lack of democratic legitimacy. Since global institutions are not constituted through democratic elections nor do they follow demo-

---

[156] Amstutz (2008, p. 217).

[157] Amstutz (2008, p. 218).

cratic decision making, they suffer from a democratic deficit."[158] Each nation has its own style of governance protected by its own sovereignty. Some weaknesses are from within specific governments and can hardly be addressed from without those state governments.

There is often "fragile ties between decision makers and citizens. Robust governance presupposes a high level of social capital—that is, a high level of voluntary cooperation based on shared values, interests and trust."[159] Without what Amstutz calls social capital which is voluntariness to cooperate on common values, interest and trust, establishment of community is not possible. Thus some national states, to begin with, are not themselves a community in strict sense. Creation of international community based on their being already community would be logically absurd and counterproductive.

Centralized government presupposes some sort of community that is governed; otherwise the governance is empty of meaning.[160] Community, in turn, presupposes "shared values and interests. The authority of law depends not only on the coercive power of institutions but also on a moral-cultural consensus. Legitimate governmental authority can exist only where a strong, consensual political culture exists."[161]

One of the base factors which enable creation of global community is global common good. Among the types of global common good are public goods such as "ideas, values, practices, resources, and conditions that benefit everyone in a society or community. Global public goods are those collective goods that extend across borders. Examples of such goods include peace, financial stability, poverty reduction, clean air, environmental protection, and conservation of the species."[162] Being shared by all, global public good is like glue that facilitates bonding which is necessary for creation of global community. Amstutz observes two important characteristics of public goods: "first their enjoyment is not diluted or compromised as the good's usage is extended to others … second, no person can be excluded from enjoying a public good."[163]

One of the sources of conflict and disagreement between states is the fact that while some states work hard to protect and safeguard public goods such as the atmosphere, oceans, and soil, others do not care. They recklessly exploit them. Amstutz notes that "the extent to which states implement sustainable development strategies domestically is vitally important because domestic practices will profoundly affect transboundary air and water pollution and thus impact the quality of the earth's atmosphere and oceans as well as the prospect for long term economic growth."[164] Thus, even though there is no international community in a strict sense, there is

---

[158] Amstutz (2008, p. 220).

[159] Amstutz (2008, p. 220).

[160] Amstutz (2008, p. 220).

[161] Amstutz (2008, p. 221).

[162] Amstutz (2008, p. 222).

[163] Amstutz (2008, p. 222).

[164] Amstutz (2008, p. 223).

inevitable transnational influence and effect due to the common or public goods shared by all.

There is need to "balance national interests with global goods, or short-term needs with long-term concerns."[165] This need can only be effectively addressed if there is a real relationship between nations. However, "The international community's institutions remain politically underdeveloped. The world remains a decentralized community where states—not intergovernmental, nongovernmental, religious movements or advocacy networks—are the primary actors." Unfortunately, such non-governmental agencies are so limited by states' sovereignty that they are often rendered helpless in the face of tremendous issues like pollution that endangers all life on our planet. Amstutz states that "promoting the global common good ultimately involves cooperative action among states, especially the largest, most powerful and economically developed countries."[166]

Lack of global government leaves citizens of any particular state at the mercy of its national government. If the government is oppressive, exploitative or dictatorial, its citizens have nowhere to appeal. "The limitations of global governance are especially evident in promoting human dignity. Despite an expansion in humanitarian international law, gross human rights abuses persist, especially when ethnic and religious groups compete for political power or when regimes pursue political repression."[167]

There is need to check on the authority of individual states and how that authority is used over its people and how it affects other peoples outside its boundaries. Amstutz warns that "Until states cede more sovereignty and create institutions to make and enforce law, the international adjudication of crime will have only a marginal impact on global society."[168] Ubuntu recognizes human species' essential unity which is not only transnational but also trans-species. Human action has effect over other humans and other species and the planet. The community as a whole should see to it that individual or community action does not hurt other humans or future generations or the planet.

4.3.2 Discrimination

The second concept of diversity is that no individual or group should be discriminated against or stigmatized on the basis of uniqueness.[169] Beauchamp and Childress address the problem of human fundamental equality and the obvious unequal global access to health care as an issue of justice.[170] Among criticisms represented

---

[165] Amstutz (2008, p. 238).

[166] Amstutz (2008, p. 238).

[167] Amstutz (2008, p. 238).

[168] Amstutz (2008, p. 238).

[169] Rivard (2009, pp. 188–198).

[170] Beauchamp and Childress (2009, pp. 240–281).

by Shetty is that the UNESCO Declaration on Bioethics and Human rights discriminates against underdeveloped countries by assuming and setting the same standard
for all countries.[171] Article 11 of the UNESCO Universal Declaration on Bioethics
and Human Rights rules out any form of discrimination based on gender, age, disability or physical, mental, social conditions, diseases or genetic characteristics.

Article 11 is founded on articles 1 and 2 of the declaration, that is, all persons
are born free and equal in dignity and human rights, all persons, therefore, share
human basic freedoms.[172] Amstutz observes that "despite the divergent theories,
competing ethical and philosophical justifications and contested interpretations of
human rights, there is widespread political acceptance of the idea of human rights
in the contemporary world."[173] This global acceptance of human rights is based on
implied acceptance of a shared common human dignity.[174]

Sweet and Masciulli state that "dignity is a characteristic of humanity, and not
just of this or that human individual, that an offense against one person's dignity is
an offense against human dignity in general."[175] In an interview Jean states that one
of the greatest challenges in bioethics is to reach an equilibrium between individual
wellbeing and needs against that of the society.[176] Such equilibrium would mini-
mize discrimination. Consequently, human dignity cannot be put aside; it has to be
recognized and respected by all cultures and peoples. Nondiscrimination is based
on human common dignity.

Discrimination is based on a false assumption that certain people, cultures or traits
make one a better human being than others. UNESCO's non-discrimination policy
is founded on the principle of human equality. In Ubuntu culture discrimination is
a serious moral evil. Ubuntu utilizes difference positively following the principle
of subsidiarity, that is, difference is utilized for the good of all by division of labor
based on one's ability or disability, gender physical strength and skills for the common good.

4.3.2.1   Amstutz on Cultural Diversity and Ethics of International Human Rights

One of the greatest assumptions, one on which personal, national and international
ethics is based, is that of human rights. Based on their inherent dignity all humans
have basic rights which ought not to be violated. Amstutz notes however, that "Because the international community is a society of societies, each with its own social,

---

[171] Shetty. (2012).

[172] Rivard (2009, pp. 191–192).

[173] Amstutz (2008, p. 95).

[174] Sweet and Masciulli (2011).

[175] Sweet and Masciulli (2011, p. 9).

[176] Jean (2004, p. 5).

political, and economic institutions and cultural traditions, defining human rights and the policies likely to enhance human dignity is a daunting task."[177]

Basic human rights are inviolable in the sense that violating them would mean violating humanity itself. Occasionally, however, there are some conflicts between human rights and some cultural practices. Hence "the challenge posed by cultural pluralism is how to reconcile universal human rights claims with the fact of cultural and moral relativity."[178]

To some extent cultural diversity is possible between different cultures and the demands of human rights. However, Amstutz notes that "the claim of total cultural diversity is simply unattainable ... diversity cannot be total because certain moral principles are necessary for social life as such, irrespective of its particular form." Amstutz observes that "there is common morality shared by all peoples. This morality involves such moral norms as justice, respect for human life, fellowship, freedom from arbitrary interference and honorable treatment."[179] At the level of common morality, there are hardly any conflicts between human rights and specific cultures. "

The challenge for the international community is to delimit human rights and to emphasize only those rights considered essential to human dignity."[180] The challenge to most indigenous cultures is to discourage cultural elements which conflict with universal human rights. UNESCO Declaration on Bioethics and Human Rights respects cultural diversity while, at the same time underlines the importance of respecting human rights, based on human dignity.

Although there is an institution responsible for the reconciliation of all global cultures with universal human rights, the reconciliation is crucial. Amstutz cautions that "in reconciling cultural relativism with the universality of human rights, it is important to emphasize that universalism and relativism are not mutually exclusive categories but rather different ends of a continuum." For acceptability of the necessary adjustment on the side of specific cultures, Amstutz's caution is important. It speaks to the approach that should be adapted. Both human rights and specific cultures aim at the good of society.

"The choice is not between the extremes of radical universalism, which holds that culture plays no role in defining morality, and radical cultural relativism, which holds that culture is the only source of morality." Any approach which involves mutual exclusivity between human rights and specific cultures is bound to escalate conflicts and eventually fail. Amstutz states that "the affirmation of human rights in global society will necessarily be based on an intermediary position that recognizes both the reality of cultural pluralism and the imperative of rights claims rooted in universal morality."[181] Thus the appropriate stance is that of 'both and,' rather than that of 'either or.' Ubuntu believes deeply in the importance of diversity. Actually

---

[177] Amstutz (2008, p. 88).
[178] Amstutz (2008, p. 92).
[179] Amstutz (2008, p. 92).
[180] Amstutz (2008, p. 98).
[181] Amstutz (2008, p. 93).

according to Ubuntu diversity and otherness are necessary for self-identity and realization, humans being human because of the otherness of other humans.

4.3.2.2   Daniels and Social Obligation to Promote Preventive Health for All

Daniels' first premise in his defense for promotion of preventive health care for all is that health is the basis and condition of most opportunities in life. That being the case, "meeting health needs protects the range of opportunities people can exercise, then any social obligations we have to protect opportunity implies obligations to protect and promote health for all people."[182] Hence, in Daniels' own words, "Meeting the health needs of all persons, viewed as free and equal citizens, is of comparable and special moral importance." Moreover, Daniels consider preventive and curative healthcare to be a basic human right. Denial of healthcare, in his view, is an injustice.

The community of nations and each state has an obligation to promote and protect human health. Daniels explains, "Just health requires that we protect people's share of the normal opportunity range by treating illness when it occurs, by reducing the risks of disease and disability before they occur, and by distributing those risks equitably."[183] Daniels underlines the importance of meeting the health needs of all people fairly by making "priority-setting decisions about all these obligations through a fair, deliberative process." Daniels goes even further by arguing that "we owe people when we cannot restore their loss of functioning: our obligations take us outside the health sector."[184] This argument is based on his premise that "the special importance of health for protecting opportunity gives us social obligations to promote and protect health. To meet these obligations and to secure equity in health, we must design appropriate policies both inside and outside the health sector."[185] Daniels argument raises a lot of questions with regards to personal accountability for health. He clarifies this controversy by arguing that "Emphasizing our social obligations to meet the health needs of free and equal citizens, regardless of how those needs arise, does not mean that we cannot hold people accountable in reasonable ways for their behaviors." However, he maintains, "We must temper our judgments in light of what we know about the determinants of health and of risky behaviors, and where we have reasonable disagreements about what we do, we must be accountable for the reasonableness of our decisions."[186]

Promotion of healthcare for all implies a degree of intrusion into personal autonomy and behaviors. Some personal preferences may have to be restricted for the sake of the health of others. Efforts to respond to a threat of spread of infectious disease, for example, "raise difficult questions about the appropriateness of restricting individual choices to safeguard other people's welfare." Examples include the

---

[182] Daniels (2008, p. 141).
[183] Daniels (2008, p. 141).
[184] Daniels (2008, p. 157).
[185] Daniels (2008, p. 157).
[186] Daniels (2008, p. 158).

use of isolation and quarantine for tuberculosis and pandemic influenza.[187] Taking responsibility for the health of others ought to a reasonable degree, limit individual autonomy.

The extent to which this kind of restriction can be imposed is a philosophically difficulty issue to discern. It may go as far as public restrictions on habits such as smoking, poor diet or lack of exercise. From the global perspective, "defining the scope of countries' obligations to act collectively, and determining how those obligations should be enforced, will inevitably raise difficult ethical dilemmas."[188] However, in line with Ubuntu world view, no human person can claim to be completely free from responsibility for other humans. A person is a product of many interpersonal relationships; disentangling a person from other persons is tantamount to annihilating him. Each human is to an extent responsible for the entire human species.

4.3.2.3 Petrini and Gainotti on Personalist Approach to Public-Health Ethics

Petrini and Gainotti observe that "The principle of autonomy has tended to dominate healthcare ethics especially in North America." In their view the dominance of autonomy in healthcare may not always be to the advantage of healthcare since, they argue, "public health is based predominantly on population-level utility, making it more attentive to issues such as epidemics, social determinants of health, and cost-effective decision making." Petrini and Gainotti admit that "a pervasive utilitarian component in public health is thereby undeniable." Petrini argues against the philosophical idea that public health is paternalistic, especially because it involves states' intrusion into personal liberties for the sake of promotion of health and safety. In their view, "The main challenge lies embedded in the relationship between individual and population health."[189]

Petrini and Gainotti contend that "If we want to promote development from a health viewpoint, we must move from a solitary, individualistic approach to a Personalist approach in an integral sense." Petrini and Gainotti believe that individualistic approach to healthcare is an impediment to real progress, hence "Going forward, we must rethink the concept of coexistence in our world, starting from the assumption that we all belong to the human species, with consideration of our different identities and, therefore, shift from the 'individual' to the 'person.'"[190] According to Petrini and Gainotti, "the founding basis of universalism, personalism and solidarity as an anthropological concept is shared, today, by representatives of different cultures."[191]

Petrini argues that "Personalism, which suggests building up the common good on the basis of attention to and care for the good of each person," is the best way to solve

---

[187] Coleman et al. (2008, p. 578).

[188] Coleman et al. (2008, p. 578).

[189] Petrini and Gainotti (2008, p. 191).

[190] Petrini and Gainotti (2008, p. 627).

[191] Petrini and Gainotti (2008, p. 627).

conflicts between individual interests and social interests.[192] In his view personalism is what is what is lacking in modern medicine, absence of which accounts for most ethical social conflicts. In Petrini's view "personalism is the best approach to face ethical problems not only in clinical bioethics, but also in public health ethics."[193]

Personalism defends public health approach to medicine. Public health is well defined by the Institute of Medicine (IOM) as "what we, as a society, do collectively to assure the conditions for people to be."[194] Total embrace of public health would imply that it is unethical to exclude anybody from healthcare, regardless affordability argument. Thus "Public health practice is characterized by global attention to whole populations and therefore by an emphasis on collective health conditions, prevention, and social, economic, and demographic determinants of health and disease."[195] Personalism, which Petrini advocates, is a form of communitarian ethics since it "rejects the notion of timeless, universal, ethical truths based on reason." Personalism recognizes the role of reason in morality but also recognize the significant role of human relationship and community.

Like it is the case with ethics of care communitarian theories consider morality to be cultural concrete and relational rather than abstract, rational and indifferent to human relationship. "Communitarians maintain that our moral thinking has its origins in the historical traditions of particular communities. Communities are not simply collections of individuals: they are groups of individuals who share values, customs, institutions, and interests."

In other words, abstracting ethical theories from their rightful human relationships and interconnectedness is in itself unethical. Petrini posits that what is "communitarian seeks to promote the common good in terms of shared values, ideals, and goals. In the communitarian perspective, the health of the public is one of those shared values: reducing disease, saving lives, and promoting good health are shared values."[196] The unity of human species evident in personalism is the same unity that the ideal of Ubuntu aspires. There is, therefore, a lot in common between the ideal vision of Petrini and Gainotti in personalism and the Ubuntu worldview.

4.4   Biosphere

Another important component of the UNESCO Universal Declaration on Bioethics and Human Rights is respect for the biosphere as the cosmic context for discourse on ethical responsibility. This component consists of two important concepts. First, all humans have an ethical obligation towards other forms of life and the cosmos.

---

[192] Petrini (2010, p. 197).

[193] Petrini (2010, p. 197).

[194] Petrini (2010, p. 189).

[195] Petrini (2010, p.189).

[196] Petrini (2010, pp. 192–193).

Second, life sciences have a duty to respect and preserve genetic integrity of both human and non-human generations.

4.4.1 Ecological Environment

The concept of being sensitive to the biosphere implies that every human individual and society has an ethical duty to protect other forms of life, the biosphere and bio-diversity.[197] Article 17 is concerned with protection of the environment, biosphere and biodiversity as a human ethical responsibility. Allison warns against limiting bioethics to a 'doctor-patient' relationship, and argues that human relationship with animals and the environment in general is within the subject matter of bioethics.[198]

In drafting the Universal Declaration on Bioethics and Human rights the UNES-CO distributed questionnaires regarding its content. Macpherson notes that 60% of the respondents to the questionnaires "wanted the scope of the declaration to encompass all life forms, not just human life."[199] The use of biotechnology should help resolve human predicaments and promote prosperity without hurting other forms of life and the cosmos.

Human activity has not always been sensitive to its negative impact on the environment.[200] Amstutz laments the absence of central global authority to regulate national/state impact on the environment. He explores the harm caused by different national states as a matter of justice, thereby unveiling the underlying need for protection of what he calls the 'global commons.'[201]

Protection of the biosphere and other forms of life is one of the major concerns of UNESCO due to the problem of extinction of some species and environmental pollution resulting from human activity. The culture of Ubuntu has always been protective of other forms of life and the environment. Ubuntu recognizes interactive and interdependent relationship between humans and the biosphere. Killing of animals except for food or in self-defense, setting unneeded fires, or cutting trees is considered an ethical evil. Respect for other forms of life and the environment is almost a religious devotion. Violence to the environment leads to violence against humans.

4.4.1.1 Faunce on Technology, Health Care, Environmental Ethics and Rights

Empirical studies show that there is real interaction, cause-effect relationship and mutuality between technology, health care, human rights and environmental ethics. Faunce relates that the "intersections between international human rights, health

---

[197] Tandon (2009, pp. 248–254).

[198] Allison (2005).

[199] Cox Macpherson (2007, p. 588).

[200] Tandon (2009, pp. 247–253).

[201] Amstutz (2008, pp. 167–196).

care and environmental ethics on the one hand, and international trade law on the other, provide one of the great normative challenges for global health policy as we emerge from the era of corporate globalization."[202] After the World War II there was global recognition of human rights due to its dignity. This move was a reaction to the abuse of the war against human dignity. Faunce, however, laments that three things were marginalized about the normative content of societal impact of global health care ethics and rights:

> The first involved how ethics and law could protect the role of the environment in human health as well as its intrinsic value to the health of all life forms. The second concerned the expanding influence of international trade law in shaping influential normative systems largely unresponsive to health care (or environmental) ethics and rights. The third concerned how emerging technologies should be regulated to help resolve some of the great problems facing humanity and its environment.[203]

Human right to health has "often been interpreted as a largely symbolic, non-enforceable individually, progressively realizable concession to normative decency or attempt to claim political legitimacy."[204] There has been an increased awareness of "justifiable and enforceable international human rights as part of any functional social contract" governing how humans treat each other regardless governmental influence and control; "Article 12 of the ICESCR importantly in this context created an international right to health, legally binding those parties who have ratified it."[205]

This involves "core obligations to provide the basic preconditions for existence, including food, water, fuel, sanitation, housing, reasonable access to essential health services and products as well as capacity to live in non-toxic environment."[206] What is regrettably missing as Faunce rightly notes is "consideration of how human beings should make basic rules governing their relationship with the environment including how new technologies should be responsive to its sustainability."[207]

Faunce foresees a great possibility of development in such a way that "norms of international human rights, bioethics, medical and environmental ethics are likely to play important roles in developing any new global social contract." All those factors, in Faunce's view, might combine to "support the concept of global public goods" in such a way that "no individual or ecosystem should be excluded." Some examples of how this strategy could be implemented include "emerging technologies facilitating clean air, equitable access to food and energy, peaceful societies, control of communicable disease, transport and law and order infrastructure, as well as sustainable ecosystem.

Related global public goods will require international cooperation for their production." Faunce argues that as global awareness levels increases about the plight of the poor populations of the world and as credible and accessible data accumulates,

---

[202] Faunce (2011, p. 49).
[203] Faunce (2011, p. 50).
[204] Faunce (2011, p. 55).
[205] Faunce (2011, p. 51).
[206] Faunce (2011, p. 55).
[207] Faunce (2011, p. 51).

"it will no longer be acceptable in health policy debates to rationalize widespread deaths among increasing numbers of poor, uninsured patients and those who cannot obtain access to essential medicine or other valuable new health technologies."[208]

Just as important is the development of global legal system that oversees and ascertains just treatment based on human equality and equity between humans but which is related to the development and use of new technologies that will not exclude or marginalize portions of human population and that will put into consideration environmental sustainability:

> When sixty three experts, for example were asked to specify which aspects of nanotechnology could most assist the developing world, the nanotechnologies cited as likely to be important in this context were nanomembranes for water purification, desalination and detoxification, nanosensors for the detection of contaminants and pathogens, nanoporous zeolites, polymers and attapulgite clays for water purification, magnetic nanoparticles for water treatment and remediation and TiO_2 nanoparticles for the catalytic degradation of water pollutants.[209]

Faunce concludes that "both international human rights and global health care ethics carry the promise of enlarging the objects of human sympathy and so the applicable range of foundational virtues, principles and rules available to decision makers." Faunce's optimism is healthy because of its holistic and productive promise that tend' to address the major global ethical issues simultaneously. To underline the importance urgency of his argument Faunce states that "foundational environmental virtues, such as "sustainability" and "solidarity with endangered species and habitats" respecting the earth itself as a self-sustaining entity, must now begin in academic and policy discourse to take their place alongside "justice" and "equality" in health care debates about the wise use of emerging technologies."[210]

Faunce's perspective is plausible not only because of its realistic grasp of the holistic integral and interrelational nature of cosmic reality and human species but, especially, because of the urgency on the part of the human species to play their rightful role of stewardship.

There are a number of disturbing facts that underline the urgency of Faunce's perspective. There is even now undeniable evidence of human failure to ascertain good stewardship not only for the planet earth but also for fellow members of the human species:

> Particular challenges for the global health care ethics and human rights in the era of globalization will be the million or so women and girls under 18 trafficked annually for prostitution; the 10 million refugees; or five million internally displaced persons, the victims of any one of the 35 or so wars currently raging across the earth; of state-promoted torture or rape in the guise of 'ethnic cleansing'; or any of the 250 million children exploited for labor, sexual gratification or as soldiers. This is in addition to 1.2 billion people living in severe poverty, without adequate obstetric care, food, safe water or sanitation.[211]

---

[208] Faunce (2011, p. 58).

[209] Faunce (2011, p. 59).

[210] Faunce (2011, p. 59).

[211] Faunce (2011, p. 59).

Plausibility of Faunce's ideas cannot be doubted. The need for holistic and realistic approach to the integration of emerging technology, healthcare, environmental ethics and human rights has never been more urgent. Ubuntu worldview of interdependence of human species, the species' interdependence with its environment and importance of care for the biosphere is a basic inspiration to the direction Faunce points to. Korthals elucidates Faunce's argument with regards to the importance of human stewardship of their environment.

4.4.1.2   Korthals' Ethics of Environmental Health

There is a direct relationship of interdependence between the biosphere and human beings. Human beings' environment not only supports human existence, it influences it substantially and conditions it. Korthals explores this fact by relating environmental health and human health. Korthals lists at least four steps in the criterion of establishing unethical environmental influence on human beings. The first is "identification of what type of problem is an environmental factor causing unhealthy influences and where the problem is located."[212] Using the example of obesity Korthals demonstrates how complex it may be to identify a bad environmental influence and its location.

Obviously, if a problem is named, identified and located, a search for solution is destined to fail. The second step is "the ethics of doing research into the factors that produce environmental hazards." Definitely the research itself has to be ethical if it has to lead to ethical results. The third step is assumption of the responsibility to manage and increase the environmental health of the people involved, and the fourth step is ethically establishing the right to intervene.[213] For a demonstrative example, Korthals sites a suburban town in New York which was constructed on a former chemical waste disposal site. He mentions how the demography of its inhabitants suffered from numerous problems related with toxicity. Some of such problems are asthma, cancer, and urinary tract infection.

To demonstrate the credibility of his argument and its validity, Korthals laments how "Government scientists made many mistakes in identifying the exact causes of the health problems that these citizens had, and resisted the data and findings of citizen activists."[214] Thus some people may be forced to live in unethical environment without their knowledge and consent. Some governments and organizations may be a bad influence on the environment of some people, in which case the innocent citizens are forcefully victimized. Korthals provides an example to demonstrate how tricky it may be. He writes,

> When on-street eateries such as McDonalds, KFC, Fish'n'chips and Ben and Jerry's are tolerated not only in cities, but also in mass media advertising and sponsorship, it should not be surprising that the numbers of obese persons are greatly increasing, as they still are in

---

[212] Korthals (2011, p. 425).

[213] Korthals (2011, p. 425).

[214] Korthals (2011, p. 415).

Britain, along with increased instances of concomitant diseases, such as type-two diabetes, cancer of the intestines and cardiovascular diseases.[215]

Clearly masses of people who are poor or uneducated are forced by their environment to eat unhealthy foods and face the consequences. The environment, which disguise as friendly; it rather uses them as means to making money regardless of their wellbeing.

Citing World Health Organization (WHO) report published in 2004, Korthals explains how our commercialized environment of plenty works against our own good. "We live in an obesogenic environment" because many people take foods which are so rich in calories that the proportion between energy in-take and energy out-put is disproportional. Some foods are too rich in energy while human physical activity has reduced.[216] Unfortunately, efforts being made to reduce obesity have been undermined, sometimes on purpose, and even used for economic gains, thus reducing the obese into means to economic gain. This fact is easily demonstrated by Korthals' example.

The American Obesity Association (AOA) formed in 1995, is nominally "a lay advocacy group representing the interest of the 70–80 million obese American women and children and adults afflicted with the disease of obesity." However, the Association "receives most of its funding—several hundred thousand dollars in all—from pharmaceutical industry, including Interneuron, American Home Products, Roche Laboratories, Knoll Pharmaceuticals Ltd., and Servier—all of which market or develop diet pills."[217]

In sum, there is no doubt, therefore, that obesity, like some other diseases, is multifactorial in origin. It can be partially genetic, overeating, or eating of unhealthy foods, either by choice or by organized or unorganized force. Since obesity is a disease it is an ethical issue. To the degree it is caused by human beings, human organizations directly or indirectly, it deserves ethical attention and analysis.

Citing Minkler's "Personal Responsibility for Health: Contexts and Controversies," Korthals argues that "improving environmental health requires attributing responsibility to people, institutions, networks and policy agent, which is often connected with differences in power and interests."[218] The environment a human subject finds himself has a huge impact on his life and health, including his self-actualization and happiness. Ubuntu world view espouse human environment and emphasize its significance, not without reason.

4.4.1.3 Tandon on Protection of the Environment, the Biosphere and Biodiversity

It is undeniable fact that the human species is sustained in existence by in interaction of many other member creatures of the planet earth. The human race can by

---

[215] Korthals (2011, p. 414).

[216] Korthals (2011, p. 417).

[217] Korthals (2011, p. 418). In this citation Korthals based his data on CSPI 2003, p. 17.

[218] Korthals (2011, p. 416).

no means survive independent of the biospheric environment which is a network of many organic and inorganic beings. Article 17 of the UNESCO Declaration on Bioethics and Human Rights states "Due regard is to be given to the interaction between human beings and other forms of life, to the importance of appropriate access and utilization of biological and genetic resources, to respect for traditional knowledge and to the role of human beings in the protection of the environment, the biosphere and biodiversity."[219] Superior as it may be to other living and nonliving beings; human species is contingent and deeply dependent and sustained by its lower living and nonliving part of planet earth.

Tandon reminds us of an important fact about human beings' relationship with the planet earth and its ecosystems. He states that, "The earth system consists of physical and biotic components, which have evolved together in continuous interaction towards its present state of complexity." In other words, independent of human activity, the earth system has been sustaining itself by keeping the healthy balance it needed at any particular time in its on-going evolution. Tandon notes that "Over the past few decades, scientific work has established that human activities have caused abrupt and unprecedented modifications in the planetary life-support system."

It is important to carefully discern whether such changes are for the good of the planet and its life forms—therefore for the human species—or not. This is why bioethics is essential. Any harm done to any component of the holistic nature of the planet affects not just that part but also all other parts, including humans and future generations. Tandon names the component parts as "the atmosphere, the marine and the terrestrial compartments." All three function together in self-sustaining synergy which Tandon calls "fluxes of matter, that is the hydrological and the biogeochemical cycles. The earth system, is in principle one and indivisible, because all parts are interconnected by delicate control mechanisms operating on various space and time scales."[220] It is the planet Earth's automated and self-sustaining principle that calls caution to human interaction and its effect over all the system.

There has been a notable change in the earth life systems due to human intervention recently because of the "advent of the industrial revolution, the development of the chemical industry and the introduction of nuclear technology." Atmospheric pollution, soil pollution and water bodies pollution, along with human over population of some earth parts has already been proven to be hazardous to some species. Consequently Tandon warns that "recent advances in molecular biology, recombinant technology, genetics and biotechnology" should be vigilantly monitored by public system to "prevent adverse effects on the environment."

Just as living and nonliving organisms' relationship among themselves and between each other is complex and interdependent in many complex ways, so is evolution. Tandon states that "scientific disciplines such as biology, sociology and economics show us that our evolution involves not only competition for survival of the fittest, but a high degree of collaboration (symbiosis) for the survival of the global living system." Needless to mention, human rationality and free will that enables

---

[219] Tandon (2009, p. 247).

[220] Tandon (2009, p. 150).

him to effect substantial change, even annihilate the planet as we know it, must be controlled and carefully utilized. "The new development of technologies must therefore respect local and national social, cultural and environmental constraints, and should pose no risk of irreversible damage."[221]

Since there have already been adverse effects on the planet and its life systems, "Environmental security is no longer peripheral to the issues of human health, food and nutritional security. It is an integral part of it and neglecting it yesterday has proven costly today, and could prove far costlier tomorrow." The most important tool needed for the care of the planet is knowledge. Understanding of the many ways human activity changes the planet and the life in it is crucial. Tandon observes that "it has been well recognized that no valid socio-economic or technological paradigm can be built unless man's relationship with the ecosystem and the universe is properly understood and cared for."

Due to humans' evolving understanding of the ways the planet sustains itself and the life it contains there is need for holistic approach. Tandon cautions that "This holistic paradigm demands a technology with a human face, used as an instrument to serve both humankind and nature. The world needs to manage itself as a system" regardless of human ability to manage it with his limited understanding.[222]

Bioethics "is concerned with the moral relevance of human intervention in relation to life. In its broadest sense it is concerned with all life forms: plants, animals including humans, and the diverse ecosystems."[223] The main concern of bioethics is to caution and to ascertain healthy relationship not just between humans' treatment of each other, but especially humans' treatment of the other forms of life and the planet earth which sustains that life. Thus bioethics cannot ignore its duty towards the cosmos and its contents. Doing so may hurt human species irreversibly.

"The inescapable fact is that the introduction of new technologies necessary for development brings with it irreversible social, ecological, and health consequences, which under certain circumstances can be harmful."[224] It is because of this possibility of harm that bioethics should be concerned with the relationship between humans and their environment. The harm humans may inflict on the planet and its life forms "must be anticipated, recognized, prevented and mitigated if we are to avoid disaster of the kind most developing and developed countries are facing today."[225]

This noble task of bioethics is much more basic than its duty in discerning morality of human treatment of one another. Its importance springs from its foundational and essential nature. Humans, however, are "an integral part of the biosphere has responsibilities and obligations towards all other forms of life."[226] Needless to say, humans' responsibility towards the biosphere, hydrosphere and the earth generally ought to be one of stewardship. Ubuntu worldview which endears, cherish and nur-

---

[221] Tandon (2009, p. 251).

[222] Tandon (2009, p. 253).

[223] Tandon (2009, p. 247).

[224] Tandon (2009, p. 247).

[225] Tandon (2009, p. 247).

[226] Tandon (2009, p. 248).

ture the cosmos and the beings in it almost as fellows is a great inspiration to the attitude that is needed in human relationship with the cosmos.

4.4.2   Future Generations

Humans have an ethical obligation to the biosphere. Life sciences have an ethical obligation to safeguard future generations, including their genetic constitution.[227] Article 16 of the UNESCO Universal Declaration on Bioethics and Human Rights posits that bioethical issues should be considered, not just for the present generations but also for the future generations. Present decisions affect future generations.[228] Because of the Declaration's position with regards to minority, especially its position against abortion, it has won support of religious groups including the Vatican.[229]

Some critics argue that the declaration is minimalistic and vague because of its failure to be specific with regards to the use of language that is too general or unclear—phrases like "impact of life sciences on future generations...should be given due regard" are harmful to the message of the document.[230] Using the language of Benatar, that kind of statement "gives guidance where none is needed and it fails to give guidance where it is needed."[231] On the contrary, some scholars like Langlois establish the relevance of the declaration while emphasizing the role of contextualization of the general principles.[232] Thus, present generations are responsible for their actions that impact future generations.

Gene therapy and human genome information, for example, may provide accurate diagnoses and therapies for individuals but may also involve serious adverse consequences for the next generations. Allison argues that the present generation has ethical "duties" to future generations.[233] Taylor suggests both national and global cooperation in benefiting from genetic research without violating human rights.[234] Morisaki suggests involvement of many parties in the decision making process as a way of regulating reckless or inconsiderate, harmful steps.[235]

It has been demonstrated in Chap. 2 how Ubuntu respects and protects integrity of both human and non-human lives of the present and future generations. In Ubuntu mindset, destruction of the integrity of future generations means, at the same time, self-destruction. In Ubuntu culture genealogy is important because it

---

[227] Morisaki (2009).

[228] Serour and Ragab (2011, pp. 147–148).

[229] Dickson (2005).

[230] Benatar (2005, pp. 1471–8847).

[231] Benatar (2005, pp. 1471–8847).

[232] Langlois (2008, pp. 39–51).

[233] Allison (2005).

[234] Taylor (1999, pp. 475–541).

[235] Morisaki (2009).

is an essential part of self-identity and belonging. It also communicates a sense of sacred obligation to extend the genealogical line with its integrity. Such a mindset serves as an inspiration to counter modern trends and temptations to tamper with human and non-human genetic constitutions.

4.4.2.1 Ethics of Genetic Manipulation in Relation to Future Generations

One of the greatest discoveries of the nineteenth century was Gregor Johann Mendel's laws of heredity. The discoveries facilitated understanding of the origins, significance and mechanisms of genetic diversity. "The principal phenomena involved are segregation, mutation, and recombination of genes. Together these three actions, through the opportunities they generate for genetic diversity, have since been used to improve plants, animals, and micro-organisms of interest to agriculture, industry and medicine."[236]

There has been rapid development in the understanding of human genome and how this understanding could be well utilized in medicine for the good of the human species. "Now that the Human Genome Project (HGP) is an ongoing and rapidly progressing reality, and human genetic engineering is expected to become procedure, the inevitable question is how these procedures will be applied." There has been a number of ethical concerns with regards to the possible application of the knowledge and possibilities that come along with accessibility to human genome. According to Walters the use of germ line gene therapy falls into three major categories: "(1) its potential clinical risks, (2) the broader concern of changing the gene pool, the genetic inheritance of the human population, and (3) social dangers."[237]

Eugenics is one of the most feared applications of the Human Genome Project. According to Agius and Busuttil this kind of eugenics "is often looked upon as positive eugenics, directed perhaps, towards achieving human beings endowed with optimal characteristics of physical strength and beauty, intellectual genius and longevity."[238] However, the more basic question is whether our limited knowledge may interfere with natural evolution process which has developed for millions of years.

Even if there is a possibility that the present generation can make an immense contribution to the good of future generations by modifying the present genes, the risk is incalculable. Agius and Busuttil argue that the present generation has a duty to "guard the present gene pool and ensure, in the most cautious and enlightened way possible, that nothing is done which may be detrimental to future generations, and that necessary measures are taken to implement any positive measures for its enhancement."[239] Even with this caution, however, it is impossible to be absolutely certain that the germ line gene change that is introduced is in both short and long run be beneficial to future generations.

---

[236] Agius and Busuttil (1998, p. 2).

[237] Walters (1986).

[238] Agius and Busuttil (1998, p. 3).

[239] Agius and Busuttil (1998, p. 4).

Agius and Busuttil acknowledge that there is "fear of the unknown" with regards to the possible outcome of the "powerful technology in the hands of scientists." Although they, themselves support positive eugenics, Agius and Busuttil acknowledge that they "hear warnings of another impending calamity (due to the misuse of genetic engineering in human germ cells) posing a threat to the human genome of future generations unless action is taken to prevent it."[240]

The fear is well founded because as Agius and Busuttil themselves acknowledge "There is of course the immense and probably insoluble problem of determining which human characteristics, among nature's rich and superb diversity, can be improved and what constitutes the hypothetical physical and intellectual excellence that one might envisage and enhance."[241] This being the case, there is need to proceed with a lot of caution and certainty or not to proceed at all. Humans are now holding in their own hands the fate of their own species. They can easily end it as it currently is. Ubuntu respect of the sanctity of human life and its sacredness would not easily permit any uncertain manipulation. Since human morality in Ubuntu is determined by the presence of "an-other" and the way the "other" is treated, the present generation's morality is measured by its sense of stewardship for the future generations.

4.4.2.2   Kalfoglou on Reprogenetics

One of the most controversial topics discussed with regards to future generations springs from the advancement in genetic technology. Genetic research and technology originally was meant for proactive preventive and therapeutic of genetic diseases. However, as Kalfoglou rightly states, "Genetic testing can now influence reproductive decisions prior to conception, prior to the transfer of embryos into a woman's uterus and during pregnancy." Thus, even though the "original goal of most of this testing was to give couples at risk of passing a serious genetic disease on to their children more reproductive choices," clearly in practice the use has "expanded to include screening for risk of adult-onset diseases and the ability to select for socially desirable traits, such as sex."[242]

This expansion is potentially the beginning of a moral slippery slope into dangerous irredeemable situations. The beginning point lies in the fact that a human being has the ability and possibly the freedom to decide how he would want the another human being to be like regardless what is naturally right or the care recipient's right of self-determination or the long run effect on the process of natural selection.

In the process of getting the *right* or *desired* person, several embryos may be destroyed or used as mere means for the desired one. Because of this moral dilemma, some governments such as Italy have passed stringent rules to regulate in vitro fertilization due to embryo destruction.[243] Related to the moral problem of embryo

---

[240] Agius and Busuttil (1998, p. 10).

[241] Agius and Busuttil (1998, p. 3).

[242] Kalfoglou (2011, p. 179).

[243] Kalfoglou (2011, p. 185).

destruction is the problem of creation of 'savior siblings' because the savior sibling is a replacement or a means used to make present the dying child.

Usually savior siblings represent parents' selfish desire to still have the dying child after death instead of another child. A savior child is not loved and accepted for itself but for the dead child.[244] Needless to mention, the other controversial issue is that of harm (emotional, psychological, spiritual, social, economic and physical) to mothers and children. Harm usually results from the technology employed. For instance "there are short-term risks for any woman who undergoes oocyte stimulation and retrieval, including hyperstimulation syndrome, which can be a serious complication." Currently, there is not enough knowledge about the long-term effects of such procedures, especially when repeated several times.[245]

Reprogenetics may easily compromise human dignity, hence corroding the very core of all ethical principles and morality. Children may be reduced to mere commodities, humans may be reduced to a work of art designed by other humans, and the conflict between those who would like to have the best selection of traits for their siblings and those who would rather let nature decide the future of their children. This situation may lead to a great moral scenario where the child designed is denied important human functions and qualities.

A typical example is provided by Kalfoglou in the case of a "lesbian couple who were both deaf and sought out a sperm donor who had five generations of deafness in his family in the hope that their child would also be deaf."[246] Nobody currently knows the long term consequences of such selections. Many nations such as Germany, Norway, Australia, and Switzerland and some U.S.A. states have passed "Laws banning the use of any type of selection based on genetics, including the use of pre-implantation genetic diagnosis (PGD) to avoid genetic disease."[247] Genetic testing and selection, potentially good as it may seem, it may cause stigma, discrimination and marginalization of those known to have a genetic disease or disability or a trait that falls short of preferred trait. Already some people have been denied employment positions or insurance coverage.[248] Discrimination therefore is a potential problem.

Attempting to manipulate human nature to improve it may not only be playing God but may actually lead to a disaster owing to limited knowledge that humans have about their own nature and anything for that matter.[249] Other ethical concerns include unpredictable racial, gender or even trait imbalance. There may be an increase in the rate of abortions since some ambitious parents who end up not getting the traits they want in a child may opt for abortion.[250]

---

[244] Kalfoglou (2011, p. 186).

[245] Kalfoglou (2011, p. 185).

[246] Kalfoglou (2011, p. 187). Kolfoglou was citing Driscoll's "Why We Chose Deafness for Our Children," *Sunday Times,* London, April 14.

[247] Kalfoglou (2011, p. 186).

[248] Kalfoglou (2011, p. 187–188).

[249] Kalfoglou (2011, p. 188–189).

[250] Kalfoglou (2011, p. 189–190).

Human dignity of children is logically being compromised in the process. Kalfoglou states "If these technologies are used to alter the characteristics of children, there could be subtle but profound effects on how parents and society view children. If children are more a product of our desires rather than a begotten gift from God, our expectations for our children may change."[251] Thus, genetic technologies, promising as they may be, especially with regards to proactive preventive medicine, they can lead to serious negative social, psychological, demographic, emotional, economic, ethical, religious and dignity consequences. In the respect Ubuntu has for human life and how nature brings forth a new member of the society as it finds fit after its experience of an unknown time span, we find both a caution and inspiration to proceed with caution in the subject of Reprogenetics.

4.4.2.3   Morisaki on Protection of Future Generations

Article 16 of the UNESCO Declaration on Bioethics and Human Rights is about protection of future generations. It specifically addresses the "impact of life sciences on future generations, including on their genetic constitution."[252] Morisaki reports that "during the drafting process and the discussions of various draft texts, there has been consensus that bioethical issues should be considered not only for the present generation but also for future generations."[253]

Human responsibility towards the biosphere "should extend to future generations and the actual decisions taken should keep that in mind." Present generation has responsibility for future generations because decisions made by the present generation affect lives of future generations. "This implies that the concept of intergenerational justice is now at the fore of today's international environmental concerns."[254] Acceptance of the concept of intergenerational justice implies responsibility on the part of the present generation. It also implies culpability for the wrong decisions made on behalf of the future generations with relation to their genetic constitution or their environment. Takayuki states, "Humanity is not only the international community, including all people living today, but it refers to the chain of generations who collectively form one community whether living now or in the future."[255]

The importance of ethical concern for the future generations is heightened by the rapid development in technology and the easiness of effecting environmental or genetic germ line change. There is need to ethically weigh the pros and cons of decisions made for the present generation on future generations. Human genome information, for example "will provide not only accurate, personalized or individual diagnosis, but also will provide a better choice of therapeutic procedures. However,

---

[251] Kalfoglou (2011, p. 191).

[252] Morisaki (2009, p. 243).

[253] Morisaki (2009, p. 243).

[254] Morisaki (2009, p. 244).

[255] Morisaki (2009, p. 244).

such new technology may result in undesired outcomes for the next generation …
as in the case of gene therapy targeting germ line cells."[256]

The excitement of finding a solution for a health problem may easily overshadow
the implication to the future generations. To avoid this threat Takayuki recommends
that "scientists coming from the health arena should not be the only ones involved
in the decision-making process; social scientists or lay persons should also be called
upon to make a contribution."[257] This inclusion implies that bioethics committees
should not only be representative of demographically, they should "play an important role in the decision-making process." Takayuki recommends inclusion of
"multidisciplinary discussions and international co-operation, including UNESCO
activity" for the sake of reaching objectively ethical decisions in matters that affect
future generations.[258]

Analogically, future generations may be considered as children or embryos because of their inability to participate in the process of decision making which affects
them. "It goes without saying that all research involving their participation must be
subject to rigorous evaluation, monitoring and governance" as a matter of justice.[259]
Unfortunately, children have not always been protected. "Research shows that children have been victims of unethical research practices … The smallpox vaccine, for
example, was first tested on the children of researchers and then on children living
in an almshouse."[260]

The *Belmont Report* clearly underlines "protection of vulnerable persons from
exploitation in research." For the sake of justice, owing to the fact that children
cannot make informed consent, the *Belmont Report* maintains that "in some circumstances it may be fair to give preference to the participation of adults rather than
conducting research on children."[261]

Clearly, ethically children are considered "vulnerable and their inclusion must
be balanced with the need to protect them from potential harm, making the issues
of consent of parents or legal representatives, the assent of the child and the assessment of the risks and benefits particularly important."[262] However, there is a delicate balance since some research must include children and may be for the benefit
of children: "International norms tend to balance the protection of children with the
need to include them in research."[263]

The *Declaration of Helsinki* includes children among the vulnerable and stipulates that two conditions must be met before involving them in research: "(1) the
research must be indispensable to promote the health of the pediatric population;

---

[256] Morisaki (2009, p. 244).

[257] Morisaki (2009, pp. 244–245).

[258] Morisaki (2009, p. 245).

[259] Samuel et al. (2011, p. 261).

[260] Samuel et al. (2011, p. 262).

[261] Samuel et al. (2011, p. 263).

[262] Samuel et al. (2011, p. 266).

[263] Samuel et al. (2011, p. 274).

and (2) it cannot be conducted on persons incapable of providing consent."[264] On the part of children most consent may be provided by parents. Samuel, Coppers and Award state that "Ethical guidelines governing research with children should be clarified to ensure that researchers respect the rights of parents and children in the context of research."[265] The ethical concern for future generations and the need to act on their behalf as a matter of justice belongs to the kernel of Ubuntu worldview.

4.5 Conclusion

As a worldview and philosophy of life, Ubuntu ascertains human dignity and a personal freedom which meets its limits only in the freedom of others within society as necessary conditions for morality. Human life and dignity are the greatest concern. Every member of society should do everything possible to safeguard and promote it. In line with Rawls' theory of justice and the UNESCO Declaration on Bioethics and Human Rights, Ubuntu recognizes human equality as a given, a *conditio sine qua non* of morality.

It is on this necessary condition that any morality is possible. The principle of subsidiarity is based on the essential equality of human dignity, which is non-negotiable. Since one's personhood is conditioned by, and flourishes on others' personhood, society is essential for not only socio-cognitive and moral development, it is essential for meaningful human life in general. Consequently, it is an obligation of every member of society to assure to the best of his ability the survival of the society. Doing so not only confirms the individual's existence, it facilitates both individual realization and societal prosperity.

In agreement with UNESCO Declaration on Bioethics and Human rights, Ubuntu recognizes the important role of plurality. Plurality is richness on which human society thrives. Difference is cherished because it is essential for self-recognition, a person being a person because of the otherness of other persons. Disentangling a person from all others is tantamount to annihilating him as a person. In Ubuntu otherness is as important as selfhood. Each person should be responsible not only for the self but also for the entire human species.

Every person being a product of many previous generations, every person is obliged to safeguard future generations as a matter of ethics. Future generations belong to the realm of otherness that helps define selfhood. It cannot be left aside. Ubuntu, like the UNESCO declaration on Bioethics and human rights, cares about how present human activities impact future generations. Though unknown to the present generation, future generations depend greatly on the present generation. It is a grave matter to put at risk their genetic constitution. Caution should be taken, especially because of the unknown risks, given the limits of human knowledge.

---

[264] Samuel et al. (2011, p. 264).

[265] Samuel et al. (2011, p. 275).

Since human society is in symbiosis with ecosystems, the biosphere and the cosmos, the relationship between human being and the cosmos should be one of stewardship and care. Ubuntu cherishes and endears human fellowship with the environment which makes possible human life. This is an inspiration that needs to be nurtured. UNESCO Declaration on Bioethics and Human Rights is deeply concerned about ethics of human relationship with the environment. Since there is no international government which oversees ethics of personal and national treatment of the cosmos, this aspect remains a great challenge to modern society. It is possible for one state, using nuclear weapons, to annihilate the human race as we know it. This threat should be at the top of global agenda for stewardship of the human race and the planet Earth. Unchecked national sovereignty threatens multiple nations' safety, especially when reckless or hateful regimes have nuclear capabilities.

Exploring the UNESCO Declaration on Bioethics and Human Rights reveals its great similarity of ethics, perspective and objectives with Ubuntu world view. They both recognize that human dignity is nonnegotiable, that it is to be respected and promoted; they both underline the importance of plurality and diversity for human flourishing, to be encouraged and engaged for the benefit of the entire human species; they both recognize the need for good stewardship for the genetic makeup of the future generations; they both recognize and care about good stewardship for the planet earth, especially, with regards to human dependence on it. Chapter 5 will explore Catholic socio-ethical teaching in relation to Ubuntu worldview.

Chapter 5
Catholic Social Ethics: Enlightening the Role of Solidarity in Global Bioethics

One of the major concepts of Ubuntu emphasizes the role of solidarity. Considering the Roman Catholic ethical tradition can enlighten the meaning of the role of solidarity in global bioethics. I have selected Roman Catholic Socio-ethical tradition as an example of one of the oldest and organized discourse in global bioethics and ethics of care. Even though the actual names of such discourses have seldom been used in catholic social teaching literature, the substance of the teaching by large concerns ethics of care and global bioethics.

Catholic socio-ethical tradition is substantially in agreement with the UNESCO declaration on Bioethics and Human Rights. With the exception of the faith component, the similarities between the Catholic socio-ethical tradition and the philosophy of Ubuntu are remarkable. Thus, there can be a valid, validating and mutual illuminating dialogue between the two ethics. One of the most important components of Catholic socio-ethical tradition is common good. The ethical debate on responsibility in global bioethics requires respect for common good. Another essential component of Catholic socio-ethical tradition is social cohesion. The meaning of common good requires respect for social cohesion or solidarity whereby individuals collaborate within a global social context. An equally essential component of Catholic socio-ethical tradition is based on minority empowerment. This chapter explores these three essential components of Catholic Social ethics to enlighten and justify the significance of solidarity in Ubuntu.

5.1 Common Good

The ethical debate on responsibility in global bioethics requires respect for the common good. Both UNESCO Declaration on Bioethics and Human Rights and Ethics of Care implicitly recognize the significance of common good. Ubuntu philosophy is based on common good. According to Ubuntu philosophy, anybody's humanity/humanness is validated by his recognition and treatment of other human beings. In itself, this maxim affirms the central importance of common good in Ubuntu philosophy. According to Catholic social teaching, respect for the common good is based on two premises. First, all humans are created in God's image and likeness. Humans are therefore essentially equal. Humans' equal dignity demands respect for human rights of each person. Second, the personal ethical right of self-determination should be defined and limited by common good.

5.1.1   God's Image

The concept of the common good relates to being created in God's image,[1] all humans have inherent, irreducible shared worth.[2] They are essentially spiritual and social beings.[3] Respect for human dignity and common good for all humans is based on the premise that all humans are created in God's own image and likeness.[4] This understanding of the radical dignity of human persons has serious implications. It implies categorical respect for human rights.[5] The Church's understanding of human rights, however, is different from the secular understanding. Human rights are attributes of human being as person. They are radical in the sense that they do not depend on being granted by society. They ought to be recognized as already naturally given. The society should foster and protect the values that promote genuine commitment to the growth and flourishing of all people.[6] The Church, like other institutions, is not exempt from issues pertaining to, and affecting human beings whether those issues are material or spiritual.

The Church has to involve itself with the real world, which is the context and habitat of human beings who are inseparably and at once physical, social and spiritual.[7] The culture of Ubuntu has deep respect for human dignity. Like Catholic tradition, Ubuntu holds that each human being is a unique product of many interconnections created by God and help to realize itself by other humans and the cosmos. As such, a human person is a unique beginning and end. Each person commands attention, respect and dignity worth his nature as unrepeatable unique event.

5.1.1.1   Vatican II's Joy and Hope (*Gaudium et Spes*)
                on the Mystery of Personhood

Citing sacred scriptures, *Gaudium et Spes* ponders the mystery that human beings are. Being created as images of their own creator, human beings are capable of

---

[1] Hollenbach (1979, pp. 55–59).

[2] Prokes (1996, pp. 57–73).

[3] Catechism of the Catholic Church. 1994. Vatican City: Libreria Editrice Vaticana, p. 1146.

[4] Hollenbach (1979, pp. 108–118); Himes, Michael J., and Kenneth R. Himes. 1993. *Fullness of faith: The public significance of theology*, 55–99. New York: Paulist Press.

[5] Hollenbach (1979, pp. 89–100).

[6] Donnelly (1994, pp. 124–139).

[7] Pope Leo XIII (1891, p. 42).

knowing, relating with, and loving their creator. They have been so elevated that the rest of earthly creation is entrusted to their care to be used for the glory of God.[8] Human greatness, however, is related to the cosmos. Humans are not solitary creatures. They are not meant to flourish independent of their relationship with other human beings and their environment. Human realization happens in community since they are by nature social beings.[9] The kernel of Christian message is the greatest commandment: unlimited love of God and love of neighbor as oneself. True love develops from human ability to relationality and sociability.

Over and above humans' call to a healthy, mutual fulfilling relationship with one another, they are called to "communion with God." *Gaudium et Spes* observes that "from the very circumstance of his origin, man is already invited to converse with God." There is always an essential relationship between human persons and God. "For man would not exist were he not created by God's love and constantly preserved by it and he cannot live fully according to truth unless he freely acknowledges that love and devotes himself to His Creator."[10]

Human beings need to recognize God since they are created both for themselves and for God. They are made with the ability to recognize, and tend towards God for their fulfillment since they are created "to commune with God and share His happiness."[11] Tendency towards God does not limit human freedom, it rather enlightens it by giving it meaning. True human freedom is not a license to do evil, on the contrary, "authentic freedom is an exceptional sign of the divine image within" a human being. *Gaudium et Spes* posits that "man's dignity demands that he act according to a knowing and free choice that is personally motivated and prompted from within, not under blind internal impulse or by mere external pressure."[12]

Thus, authentic freedom is essential and rightful to all human beings. In other words, any external condition that impedes any human being from freely opting for goodness, which is ultimately God, is evil and injurious to the very essence of humanity. Humans are essentially rational and free beings. The shared ability to understand and to freely make an option is constitutive of personhood. Personhood implies that humans are essentially moral beings.

The ultimate vocation of all humans is reflected and manifested in Christ. Christ is both the inspiration and revelation of God's desire for humanity and humanity's craving for God. Even though humans don't share in the divine substance like Christ. In his humanity, Christ reflects the perfect or ideal human person. *Gaudium et Spes* teaches that "Christ, the final Adam, by the revelation of the mystery of the Father and His love, fully reveals man to man himself and makes his supreme calling clear. It is not surprising, then, that in Him all the aforementioned truths find their root and attain their crown."[13] Christ's authentic obedience in freedom and

---

[8] *Gaudium et Spes* (1965).
[9] Gaudium et Spes (1965, p. 12).
[10] Gaudium et Spes (1965, p. 19).
[11] Gaudium et Spes (1965, p. 21).
[12] Gaudium et Spes (1965, p. 17).
[13] Gaudium et Spes (1965, p. 22).

truth to the will of His Father is a model of human relationship with God. Such relationship is based on authentic relationship with God, founded on truth and practiced in freedom. Basing its teaching on scriptural revelation, Vatican II documents state that Christ is "the perfect man. To the sons of Adam He restores the divine likeness which had been disfigured from the first sin onward."

However, Christ is not a mere ideal and inspirational model to be imitated. His assumption of human body and condition, not only reveals humans to themselves as children of God, but "by His incarnation the Son of God has united Himself in some fashion with every man. He worked with human hands, He thought with a human mind, acted by human choice and loved with a human heart. Born of the Virgin Mary, He has truly been made one of us, like us in all things except sin."[14] Since Christ's incarnation transcends human ability to comprehend, it remains a mystery. Humans' relationship with God is both imminent and transcendental. It cannot be exhausted and should always be respected. It reveals human transcendental nature and dignity thereby demanding for its recognition and respect.

Vatican II documents state that the teaching about the mystery of human dignity applies to all humans regardless their culture, religion, and all accidents. It is essential to the human essence. In the words of Vatican II, "All this holds true not only for Christians, but for all men of good will in whose hearts grace works in an unseen way." Christ's revelation of humanity to itself and his teaching which culminated in the paschal mystery cannot be limited. It is meant for all humans. *Gaudium et Spes* explains, "since Christ died for all men, and since the ultimate vocation of man is in fact one, and divine, we ought to believe that the Holy Spirit, in a manner known only to God, offers to every man the possibility of being associated with this paschal mystery."[15] Through the mystery of Christ and the help of the Holy Spirit every human being can cry out Abba, Father.[16]

Like Vatican II's *Gaudium et Spes* Ubuntu recognizes universal human neediness for God, God's love and elevation of human persons, and humans need for other humans for the sake of realization of their dignity and the dignity of other humans. Both Ubuntu and *Gaudium et Spes* recognize human irreducible and non-negotiable equal and God-given dignity. Consequently, Vatican II could possibly consider a person who has not yet received the Christian message, but faithful to the Ubuntu worldview, an anonymous Christian.

5.1.1.2 John Paul II's Ethical Personalism in (*Redemptor Hominis*)

In an inseparable and substantive way Ubuntu recognizes the personal right to self-determination. However, personal rights are only realizable in the context of human society. Every person needs to be able to transcend his own individual rights into the common humanity that is shared by every human being. John Paul II's teaching is

---

[14] Gaudium et Spes (1965, p. 22).

[15] Gaudium et Spes (1965, p. 22).

[16] Gaudium et Spes (1965, p. 22).

deeply rooted in Vatican II documents' philosophy, which prioritizes the interests of human race in general without ignoring or undermining human individual interest. Consequently, his teaching is ecumenical, appealing to natural law and geared towards reunification of all churches and peoples. The reason given in the documents themselves is that "Christ wills it, and Christ is the head of all humankind."[17]

Christ's incarnation is for the entire human race. Assuming human nature, Christ's will is to redeem it all, not just a section of it. Christ is the revelation of God and his love and mercy to all human beings.[18] Being created in God's own image and likeness, all human beings are invited, challenged and enabled to love God in return and to love fellow humans as God loves them and as they love themselves. John Paul II states that "Man needs to love in return to be fully human. Man cannot live without love.

Unknowingly, John Paul II almost replicates the kernel of Ubuntu philosophy when he states that only in relationship to others, and to Christ, can each man discover what it means to be human."[19] To be human, according to John Paul II is to love truly. The church should preach by deeds and words this truth about humans' need to love and be loved to all peoples of the world. The main justification of the love is the undeniable fact of human dignity, revealed not only in creation, but also in the paschal mystery.[20] The shared understanding of human dignity between Catholic social teaching and the philosophy of Ubuntu cannot be denied.

Although the church is an institution, it transcends other institutions. Citing *Gaudium et Spes*, John Paul II explains that church cannot be "identified with any political or ideological system—the Church is, rather, a sign and a safeguard of the transcendence of the human person." The church represents a dimension of humanity that the temporal political and organizational systems do not represent. The Church represents spiritual and eschatological aspects of humanity, without underrating its spacio-temporal and sociological dimensions.[21]

The Church is a permanent sign and revelation of human dignity as children of God, images of God, Co-creators and co-governors of the universe with God. In a very special way, this partnership of humanity and divinity is at its peak in the Eucharist. In the Eucharist God becomes one with human kind. The Eucharist is thus the "most important source and sign of our human dignity." Liturgy, therefore communicates human dignity. "It is in respecting the Eucharist, equally a Sacrament of Sacrifice, a Sacrament of Communion, and a Sacrament of Presence, that we show honor to the One who has valued our dignity by dying on the cross."[22]

One of the most foundational aspects of human dignity is true human freedom. However, freedom has often times been misunderstood as a permission to break natural law and God's law. When freedom is abused, human dignity is compromised. The Pope states that, "Freedom is a great gift only when we know how to

---

[17] Pope John Paul II (1979, pp. 18–20).
[18] Pope John Paul II (1979, pp. 22–24).
[19] Pope John Paul II (1979, p. 25).
[20] Pope John Paul II (1979, p. 27, 30).
[21] Pope John Paul II (1979, p. 42).
[22] Pope John Paul II (1979, pp. 78–81).

use it consciously for everything that is our true good. Christ teaches us that the best use of freedom is charity, which takes concrete form in self-giving and in service."[23] There cannot be freedom that contradicts human nature to love and be loved.

True human freedom should not cause intentional harm since freedom is oriented towards God's goodness which humans naturally desire. The freedom that is found in scientific and technological development should make human life more human, more dignified. If technology and science, therefore, do not reveal even more deeply human dignity, they cannot be called development. John Paul II's concern is the fact that some of the modern scientific and technological development has been to the detriment of authentic human progress, because it has become an impediment to true human freedom, rather than promote it.

Human life is not only uncompromisingly central, it is a sacred mystery of God's love. Even though human beings are societal and find themselves within the society "Political, ideological, and economic systems must not usurp the essential dominion that each individual man has in his own life." The sanctity of human life is transcendental. The pope posits that "Man cannot relinquish himself or the place in the visible world that belongs to him. He cannot become the slave of things, the slave of economic systems, the slave of production, the slave of his own products." The pope warns that a purely materialistic civilization is self-contradictory and self-defeating. Such civilization is doomed to self-destruct. Real civilization should not enslave any human being but rather enhance human freedom. Systems that give precedence to profits or ideology, or economic classes over human persons are against true human progress. Thus communism and pure free-market economy may be impediments to true human progress.[24]

Development in technology and civilization should be concomitant with development in ethics and morals. Any advancement which does not dignify human life is false because it results from humans and should be at the service of humanity rather than diminish it. One way to know whether scientific or technological development is worthy of humanity is to examine its consideration of human moral and spiritual progress.[25] The pope explains the dominion given to humans over nature to consist of priority of ethics over technology which should be reflected in the primacy of personhood over persons' products and the superiority of spirit over matter.[26]

To clarify the contradictions in the modern development, the pope points out the plight of the poor parts of the world. He laments, "We all know well that the areas of misery and hunger on our globe could have been made fertile in a short time, if the gigantic investments for armaments at the service of war and destruction had been changed into investments for food at the service of life."[27] In this text the Pope universalizes Ubuntu philosophy, that is, real civilization and human progress cannot be real while ignoring needy human beings. No one can be free from the plight

---

[23] Pope John Paul II (1979, p. 81).
[24] Pope John Paul II (1979, p. 50).
[25] Pope John Paul II (1979, pp. 46–47).
[26] Pope John Paul II (1979, p. 49).
[27] Pope John Paul II (1979, p. 57).

of poor and the needy of the world. Theological truths about human personhood should never be left aside in any realistic human development.

Any attempt to do so is *reductio ad absurdum*.[28] John Paul II's ethical personalism is in agreement with Ubuntu worldview, especially with regards to realistic and holistic human development. There cannot be valid development, which does not do justice to its author, human person. Development, which compromises human dignity in any particular human person, compromises the dignity of the entire human genre. Like John Paul II, Ubuntu refuses to separate human progress from humans' nature as spiritual and religious beings. For Ubuntu, every human being is a spiritual being wherever he is and whatever he is doing. His connection with God, other humans and the cosmos cannot be ignored.

5.1.1.3 John Paul II on Natural Law as Participation in God's Plan (*Veritatis Splendor*)

According to John Paul II, the Decalogue and its fulfillment in the great commandment of love for God and for fellow humans reveals human dignity, which God wants to be recognized and respected by all humans. There is neither measurement nor boundaries to the extent human dignity should be respected. Human life can only be compared to another human life, thus loving a fellow human as oneself. All commandments of the Decalogue serve the great commandment of love. The negative commandments such as: "You shall not murder; You shall not commit adultery; You shall not steal; You shall not bear false witness—express with particular force the ever urgent need to protect human life."[29]

Thus the commandments reveal "the good of the person, at the level of the many different goods which characterize his identity as a spiritual and bodily being in relationship with God, with his neighbor and with the material world."[30] Human dignity springs not only from humans being images of God and children of God, it springs from the fact that being *imago Dei* and children of God, humans have been intended and wanted by God for their own sake. John Paul II used the phrase "the only creature that God has wanted for its own sake" to describe this "singular dignity of the human person."[31]

God intends human beings for their own sake and orients them towards pursuit of perfection from, and in God self. Pursuit of perfection by a human person, though assisted by divine grace, proceeds from human free will. A human person is able to, and does make "decisions about oneself and a setting of one's own life for or against the Good, for or against the Truth, and ultimately for or against God."[32]

---

[28] Pope John Paul II (1979, p. 57, pp. 72–75).

[29] Pope John Paul II. Encyclical on the splendor of truth *Veritatis Splendor*. http://www.catholic-pages.com/documents/veritatis_splendor.pdf, p. 13.

[30] Pope John Paul II (1993, p. 13).

[31] Pope John Paul II (1993, p. 13).

[32] Pope John Paul II (1993, p. 65).

Due to human dignity revealed in the commandments that God gives to safeguard it, Sin is not only an offence against God, as it is equally an offence against the sinner and the entire human species. Consequently John Paul II writes that "God is offended by us only because we act contrary to our own good." This means that there are some actions which are intrinsically evil because, in themselves, they hurt human beings. The Pope argues that, just as there are inviolable human rights there are undeniably intrinsically evil acts.[33]

According to John Paul II "Jesus brings the commandments to fulfillment ... by interiorizing their demands and by bringing out their fullest meaning." That is why the commandments are not to be kept for their own sake. They are, as Jesus himself explained, for the good of humanity. Even the negative ones are meant for the good of humanity. A commandment such as "'You shall not murder' becomes a call to an attentive love which protects and promotes the life of one's neighbor. The precept prohibiting adultery becomes an invitation to a pure way of looking at others, capable of respecting the spousal meaning of the body."[34]

The ultimate revelation of human dignity is achieved in the Paschal mystery. Jesus' crucifixion, death and resurrection and the institution of the Eucharist reveal not only Christ as son of God and God's love for human beings but also the mystery of a human person. The Catholic Catechism teaches that the dignity of the human person is grounded in the affirmation that persons are created by God and bear the image and likeness of God in a way that is unique to humanity out of all creation; the "divine image is present."[35] However, John Paul II writes, "It is Christ, the last Adam, who fully discloses man to himself and unfolds his noble calling by revealing the mystery of the Father and the Father's love."[36]

In Christ's obedience to the Father, in Christ's teaching and ultimately in his crucifixion and resurrection for humanity, human beings learn God's will for them. Discipleship, therefore, is a journey into rediscovery of true human calling to God which is not isolative of others. Such a journey is made present in Jesus Christ. Jesus does not only fulfill the law, he fulfills humanity. The ideal and objective of moral life is found in the person of Christ. Humans need to conform to Christ by "holding fast to the very person of Jesus."[37]

In Jesus Christ, humanity and divinity merge into one person who is fully divine and fully human. An important aspect of incarnation is the coming together of God's law and human law as one law from different perspectives. John Paul II interpreted natural law as human rational participation in God's eternal law. The Pope states that "the moral law has its origin in God and always finds its source in him." However, the same law of God is "by virtue of natural reason, which derives

---

[33] Pope John Paul II (1993, p. 24, 96, 97, 99).

[34] Pope John Paul II (1993, p. 15).

[35] Catechism of the Catholic Church. 1993. (Approved and Promulgated in the Apostolic Letter Laetamur Magnopere by John Paul II on August 15, 1997) http://www.vatican.va/archive/ccc_css/archive/catechism/ccc_toc.htm. Accessed 14 March 2008, p. 1702.

[36] Pope John Paul II (1993, p. 2).

[37] Pope John Paul II (1993, p. 2, 19).

from human wisdom ... a properly human law."[38] Thus divine law that is revealed to human beings is discoverable by human reason because it is meant for human observance.

In observing the divine law, humans realize God's plan for them. By virtue of their rationality and free will, humans ought to observe divine law which comes to them as natural law. Since humans are not forced to observe the law they can be held responsible and accountable for their voluntary choices. "Right to the exercise of freedom ... is an inalienable requirement of the dignity of the human person."[39] At the core of the divine law is the mystery call of the human being, who is called into partnership with God. This partnership makes human life sacred, transcendent and dignified "regardless of his or her beliefs, actions, or choices." By their very nature, "persons are also characterized by reason or intelligence, freedom and autonomy, and a spiritual soul."[40] Over and above all other earthly creatures, human beings participate in God's own plan in both existential and transcendental way. No human being should be taken for granted.

In line with Catholic social ethics, Ubuntu treasures, and protects the sacredness of human beings regardless their accidental traits. The commandments of God, which find their ultimate meaning and fulfillment in the great commandment of love of God and neighbor for the good of each person can rightly be understood in, and summarized by the Ubuntu philosophy. Ubuntu unquestionably fails to separate a human person from God and from neighbor. Even though human uniqueness and its rights are clear, such uniqueness and unity is absurd if it is cut off from God and neighbor. All commandments of God are contained in the Ubuntu ethic.

5.1.2 Individual and Community

Every human being has a right to self-determination. However, this right is defined and limited by societal pursuit of common good.[41] John XXIII in his *Mater et Magistra* states clearly that, "Individual human beings are the foundation, the cause and the end of every social institution."[42] *Mater et Magistra* emphasizes the recognition of human dignity as essential for the development of proper human prosperity.[43] It is the recognition of dignity due to every human that will facilitate authentic solidarity. Proper understanding and implementation of the common good, that is, the sum total of those conditions of social living, which enable each member to fully and readily achieve his or her own realization, should be the ideal objective of the

---

[38] Pope John Paul II (1993, p. 40).

[39] *Catechism of the Catholic Church* (1993, p. 1738).

[40] *Catechism of the Catholic Church* (1993, p. 1711).

[41] Pope John XXIII. 1961. Encyclical letter on Mother and Teacher. *Mater et Magistra,* trans. William Gibbons and Committee of Catholic Scholars (p. 65). New York: Paulist Press.

[42] Pope John XXIII. 1961. Encyclical letter on Mother and Teacher. *Mater et Magistra,* p. 219.

[43] Pope John XXIII. 1961. Encyclical letter on Mother and Teacher. *Mater et Magistra,* p. 215.

society.[44] Thus distribution of the fruit of human labor in such a way that the "common good of all society will be kept inviolate" should be the *modus operandi.*[45]

The Church observes that although social institutions are like matrices which encompass and contain their members, such institutions exist because of the persons who constitute them. Their rightful place is that of a means to an end, which end is an individual human person. Common good in practice means that the society has to assure human dignity by ascertaining the decent minimum of living and of care for all. Like Catholic Church tradition, Ubuntu advocates for maintenance of the decent minimum of health for all without violating individual freedoms and right to the fruit of their labor. The society provides security and assurance of protection to its members.

5.1.2.1   Catechism of the Catholic Church on the Human Communal Character

Catholic ideal of community is found in God himself. God is community per excellence and fulfillment of all human desire for community. The Catholic Catechism states that "There is a resemblance between the union of the divine persons and the fraternity that men are to establish among themselves in truth and love. Love of neighbor is inseparable from love of God."[46] From the Blessed Trinity, humans learn that they are created not only for themselves but also for one another and for the community. Community making and maintaining, therefore belongs to the kernel of Christianity. This theme is also found in Christ's teaching as, for example, in the parable of the vine and its branches in John 15:5 and in the priestly prayer of Jesus in John "… that they may all be one" 17:22.

Hence the core message of Christianity is reconciliation with oneself; with God; and with human community. Christian reconciliation, however, is about formation of authentic community. Human beings are called to true fulfillment in community. As far as unity is concerned Ubuntu is in total agreement with Christian message except for the Christological aspect of Christianity, which is only present in Ubuntu unanimously.

Community involves stronger bonds between persons than society. In most cases humans experience themselves as member of society rather than community due to weaker bonds between them. A society is described by the Catholic Catechism as, group of persons bound together organically by a principle of unity that goes beyond each one of them. As an assembly that is at once visible and spiritual, a society endures through time: it gathers up the past and prepares for the future. By means of society, each man is established as an 'heir' and receives certain 'talents' that enrich his identity and whose fruits he must develop. He rightly owes loyalty to the communities of which he is part and respect to those in authority who have charge of the common good.[47]

---

[44] Pope John XXIII. 1961. Encyclical letter on Mother and Teacher. *Mater et Magistra*, p. 65
[45] Pope Pius XI (1931, p. 49, 57).
[46] *Catechism of the Catholic Church.* 1994. Vatican City: Libreria Editrice Vaticana, no. 1878.
[47] *Catechism of the Catholic Church,* p. 1880.

Thus, realization of the potential that each human person has is not possible independent of the human society. By implication, therefore, there is a symbiotic reciprocal relationship between individuals and human race as a society. Just as the society needs its members to exist as one; so does each member of the society need the society to realize his individuality. In agreement with Ubuntu, the Catechism categorically states, "The human person needs to live in society. Society is not for him an extraneous addition but a requirement of his nature. Through the exchange with others, mutual service and dialogue with his brethren, man develops his potential; he thus responds to his vocation."[48] A community is a kind of society whose members are interconnected and interdependent in a deeper way than in the society. It is what Christ prayed for in John 17:22, that they may all be completely one.

The importance of human society for individual realization ought not to compromise each individual's uniqueness and autonomy. Moreover, as the catechism elaborates, the human person "is and ought to be the principle, the subject and the end of all social institutions."[49] This statement implies respect for individuals and their rights, especially about their dignity, freedoms and initiative. The catechism discourages and warns against excessive intervention into individual's life by the society.

This teaching is best explained in the principle of subsidiarity which states, "a community of a higher order should not interfere in the internal life of a community of a lower order, depriving the latter of its functions, but rather should support it in case of need and help to co- ordinate its activity with the activities of the rest of society, always with a view to the common good."[50] This principle applies from personal or individual level through national level to cosmic level. In reality, the Principle of Subsidiarity is "opposed to all forms of collectivism. It sets limits for state intervention. It aims at harmonizing the relationships between individuals and societies. It tends toward the establishment of true international order."[51] In sum, the principle of subsidiarity is essential for maintenance of peace and harmony, while supporting individual realization within human society.

The Catholic Catechism encourages creation of "voluntary associations and institutions" on all levels of the society "which relate to the economic and social goals, to cultural and recreational activities, to sport, to various professions and to political affairs." Creation of such associations explicate and "express the natural tendency for human beings to associate with one another for the sake of attaining objectives that exceed individual capacities." It develops the qualities of the person, especially the sense of initiative and responsibility, and helps guarantee his rights.[52]

The teaching of the importance of human community in the Catholic Catechism is replicated in the Ubuntu worldview. The only substantial exception is Catholic's religious dimension on which human community is founded, that is, the Sacred

---

[48] *Catechism of the Catholic Church*, p. 1879.

[49] *Catechism of the Catholic Church*, p. 1881.

[50] *Catechism of the Catholic Church*, p. 1883.

[51] *Catechism of the Catholic Church*, p. 1885.

[52] *Catechism of the Catholic Church*, p. 1882.

Trinity revealed in Christ's teaching. Even though Ubuntu ideal of community and society is religious, it is not explicitly Christian. It can be argued, however, that Ubuntu worldview is a kind of anonymous Christianity due to Ubuntu worldview's remarkable resemblance to Christian Ideals in praxis.

5.1.2.2   Catechism of the Catholic Church on Human Equality in Difference

Christianity teaches that all human beings are substantially equal. Human beings are equal because they are all "created in the image of the one God and equally endowed with rational souls." Moreover, they all "have the same nature and the same origin. Redeemed by the sacrifice of Christ, all are called to participate in the same divine beatitude: all therefore enjoy an equal dignity."[53] Thus, human beings are equal in their substance, origin and destiny. The Catechism of the Catholic Church teaches, "Equality of men rests essentially on their dignity as persons and the rights that flow from it."

Unfortunately, notes the Catholic Catechism, "There exist also *sinful inequalities* that affect millions of men and women. These are an open contradiction of the Gospel." Millions of people have been marginalized by a few wealthy in many countries of the world. Since they are essentially equal to the wealthy, it is unethical to deny the poor the same essential dignity. The catechism states that, "Excessive economic and social disparity between individuals and peoples of the one human race is a source of scandal and militates against social justice, equity, human dignity, as well as social and international peace."[54] The catechism puts to words and elaborates the spirit of Ubuntu since that is exactly what Ubuntu teaches.

Due to human essential equality, "Every form of social or cultural discrimination in fundamental personal rights on the grounds of sex, race, color, social conditions, language, or religion must be curbed and eradicated as incompatible with God's design."[55] There are non-essential differences between human beings based on "age, physical abilities, intellectual or moral aptitudes, the benefits derived from social commerce, and the distribution of wealth." Each individual human has some unique gifts that are not distributed equally with other human beings.[56]

The difference in the distribution of talents "belongs to God's plan, who wills that each receive what he needs from others, and that those endowed with particular 'talents' share the benefits with those who need them. These differences encourage and often oblige persons to practice generosity, kindness, and sharing of goods; they foster foster mutual enrichment of cultures." The talents are meant both for the individual endowed with them, and for the entire human race. Inequality and diversity of endowment reflect God's desire for people to be a community, each in need of what others ought to offer. Service, which is one of the most basic requirements to inherit eternal life, means readiness and desire to avail one's talents for others.[57]

---

[53] *Catechism of the Catholic Church*, p. 1934.

[54] *Catechism of the Catholic Church*, p. 1938.

[55] *Catechism of the Catholic Church*, p. 1935.

[56] *Catechism of the Catholic Church*, p. 1936.

[57] *Catechism of the Catholic Church*, p. 1937.

All Christians, in fact, all human beings have a duty of availing themselves to others which duty is reciprocated by all that each human receives from others. The duty to avail oneself to others is a moral obligation especially towards handicapped. The catechism states, "The duty of making oneself a neighbor to others and actively serving them becomes even more urgent when it involves the disadvantaged, in whatever area this may be."

Citing the words of Christ himself, the catechism validated its statement. Jesus stated in Mathew 25:40, "As you did it to one of the least of these my brethren, you did it to me."[58] The gravity of human obligation to other humans goes to great lengths. It involves loving those who do not love us, forgiving enemies and praying for them. Citing the Word of God, the catechism asserts, "Liberation in the spirit of the Gospel is incompatible with hatred of one's enemy as a person, but not with hatred of the evil that he does as an enemy."[59]

In sum, the Catechism of the Catholic Church acknowledges humans' essential equality in their accidental differences. This acknowledgement implies mutual responsibility and service to all humans. Special obligational preference is to those who are less gifted in the society. This teaching within the Catholic Catechism and rooted in the Gospel is shared by the philosophy of Ubuntu. Human dignity cannot be compromised. The Catechism states, "Social justice can be obtained only in respecting the transcendent dignity of man. The person represents the ultimate end of society, which is ordered to him."[60] Each person should be able to contribute in accordance with the Principle of Subsidiarity. The Society, on its part, should ensure social justice by allowing "associations or individuals to obtain what is their due, according to their nature and their vocation. Social justice is linked to the common good and the exercise of authority."[61]

5.1.2.3 United States Catholic Bishops on Economic Justice for All

The central thesis of the United States Catholic Bishops on economic justice for all is the worth and sacredness of human beings. The bishops explicitly teach that "The dignity of the human person, realized in community with others, is the criterion against which all aspects of economic life must be measured." Thus, "All human beings, therefore, are ends to be served by the institutions that make up the economy, not means to be exploited for more narrowly defined goals."[62] However, the bishops note that by Government's own official definition of poverty, one in every seven people in the United States of America is poor, which means at least 33 million Americans live below what is recognized as official line of poverty.

Given this situation, the bishops choose the path of preferential option for the poor as an alternative to safeguard the sacredness and dignity of human life. They

---

[58] *Catechism of the Catholic Church*, p. 1932.

[59] *Catechism of the Catholic Church*, p. 1933.

[60] *Catechism of the Catholic Church*, p. 1929.

[61] *Catechism of the Catholic Church*, p. 1928.

[62] United States Catholic Conference (2009, p. 28).

state, "The norms of human dignity and the preferential option for the poor compel us to confront this issue with a sense of urgency. Dealing with poverty is not a luxury to which our nation can attend when it finds the time and resources. Rather, it is a moral imperative of the highest priority."[63]

According to the United States bishops, and in line with Vatican II Documents, "fulfillment of the basic needs of the poor is of the highest priority." It is a yard stick which measures the morality of an economy. It is of crucial importance for any organization or person to have a right scale of priority concerning the poor. The bishops state, "meeting fundamental human needs must come before the fulfillment of desires for luxury consumer goods, for profits not conducive to the common good, and for unnecessary military hardware."[64]

Likewise, the United States bishops prioritize inclusion, because they hold and teach that "Increasing active participation in economic life by those who are presently excluded or vulnerable is a high social priority." Being subjects, human beings are not satisfied by being merely given some basic needs to sustain their lives. They do have social needs as well. Humans also need to participate in the process of productivity, not just for material gain, but even more importantly, for self-fulfillment. This is why justice demands much more than mere providing of food, some clothing and housing. Justice "points to the need to the present situation of those unjustly discriminated against in the past." Distribution of power is an important dimension of economic distribution since it influences the very process of distribution of wealth and labor. Power and access to it should be shared by all following the principle of subsidiarity.[65] Investment in production of basic needs such as food and education should precede investment in luxurious articles of ostentation.[66]

American bishops remind the United States of the unity of both the human species and the entire planet with its biosphere and ecosystems. They argue that American Catholics belong to the universal (or Catholic) Church. By the very virtue of being Catholic or universal, they cannot dissociate themselves from the problems of other people, wherever they may be on the planet earth. They should concern themselves with "the well-being of everyone in the world."

Consequently, global problems such as third world countries' debt, starvation in some Asian, South American, Sub-Saharan and Indian countries become Americans' concern. Global environmental crisis affects all members of the human genre. There is real global economic interdependence that cannot be ignored. The bishops write, "now is the moment when all of us must confront the reality of such economic bonding and its consequences, and see it as a moment of grace—a *kairos*— that can unite all of us in a common community of the human family. We commit ourselves to this global vision." American bishops urge for global collaboration

---

[63] United States Catholic Conference (2009, p. 170, 172).
[64] United States Catholic Conference (2009, p. 90a).
[65] United States Catholic Conference (2009, p. 91b).
[66] United States Catholic Conference (2009, p. 92c).

since common good can no longer be limited to national or regional boundaries.[67] "The unfinished business of the American experiment includes the formation of new international partnerships, especially with the developing countries, based on mutual respect, cooperation, and a dedication to fundamental justice."[68] There is need to collaborate with, and strengthen "effectiveness of international agencies in addressing global problems." Unity is increasingly becoming a necessity for global justice and common good that cannot be left aside.[69]

The central message of the document of the American bishops is again effectively summarized by Ubuntu, which affirms and defines individual personal existence only by means of another. In other words, a person is known by the way he or she recognizes other person's personhood and respect it in deed. According to Ubuntu for example, a person who can find comfort and peace while aware that another person is in need of help is not human. Humanity is always associated with the ability to empathize. Clearly, the objective ideal of Ubuntu worldview is global unity of the human species within its necessary context, the cosmos for the good both of the human genre and other species within the biosphere. Human race cannot ignore its need for other forms of life and the environment.

5.2 Social Cohesion

An equally essential component of Catholic ethical tradition is social cohesion. The meaning of common good requires respect for social cohesion whereby individuals collaborate within a global social context. Social cohesion is based on two necessary concepts. The first concept concerns solidarity as necessary for existence of human community. The second concept concerns the realization that human action should be characterized by mutuality of concern for oneself and for others. Both concepts are consistent with, and belong to, the meaning of the philosophy of Ubuntu.

5.2.1 Solidarity

Crucial to the concept of social cohesion is solidarity. Solidarity is necessary for existence of human society, which, in itself, is a system of social organizations.[70] Human society is hierarchical, beginning with the natural human family and ultimately encompassing the whole human race.[71] John Finnis defines common good as "a set

---

[67] United States Catholic Conference (2009, p. 363).

[68] United States Catholic Conference (2009, p. 322).

[69] United States Catholic Conference (2009, p. 324).

[70] Mueller (1943, p. 147).

[71] Pope John Paul II. 1991. Encyclical on the hundredth year *Centesimus Annus*, p. 13.

of conditions which enables the members of a community to attain for themselves reasonable objectives, or to realize reasonably for themselves the value(s), for the sake of which they have reason to collaborate with each other in community."[72]

The Church's teaching on solidarity is based on the anthropological understanding that humans are naturally social. Common good tradition aims at promoting conditions and institutions which are necessary for human cooperation and achievement of common objectives derived from a shared vision of humanity.[73] John Paul II gives a very comprehensive understanding of solidarity from Catholic perspective. In his *Sollicitudo Res Socialis,* John Paul II emphasizes the need to view human potential and growth in moral terms. He underlines the need for interdependence and a communal sense of sharing.[74]

Human solidarity resists objectification of another person as a mere means to self. It is a challenge to view the other person as "a sharer, on a par with ourselves in the banquet of life to which all are equally invited by God."[75] A true sense of solidarity involves a willingness to sacrifice self-interest for the sake of the other.[76] *Mater et Magistra* clearly states that economic prosperity should be based not so much on the sum total of goods and wealth possessed as from the distribution of goods according to norms of justice.[77]

Solidarity is based on mutual recognition of each society member as an end irreducible to a means for any other person. True human solidarity eliminates marginalization and exploitation of any human being by another. From Ubuntu perspective, solidarity is a rule of life. A person who separates from the community is considered dead. It is solidarity which gives life and security to individuals and societies. Consequently, Ubuntu understanding of human solidarity finds endorsement in the Catholic traditional teaching.

5.2.1.1   Christianity's Essential Message of Liberation from Subhuman Conditions

Although liberation is a central theme in Christianity, Christ having died to liberate human beings from all sorts of oppression by evil, sin, and subhuman conditions, liberation theology as a movement, whose exponents are Leonardo Boff and Juan Luis Segundo, has not always been in line with orthodox Catholic theology.[78]

---

[72] John Finnis (1980, p. 155).

[73] Christiansen (1991, pp. 43–44); Hollenbach (1989, pp. 70–94).

[74] Pope John Paul II. 1987. Encyclical letter on the concern of the church for the social order *Sollicitudo Rei Socialis.* http://www.vatican.va/holy_father/john_paul_ii/encyclicals/documents/hf_jp-ii_enc_30121987_sollicitudo-rei-socialis_en.html, p. 21, 33, 38–39.

[75] Pope John Paul II. 1987. Encyclical letter on the concern of the church for the social order *Sollicitudo Rei Socialis,* p. 39; Catechism of the Catholic Church, p. 361, 953.

[76] Dorr (1989, pp. 48–49).

[77] Pope John XXIII. Encyclical letter on mother and teacher *Mater et Magistra,* p. 74.

[78] Webster (Accessed 12 May 2012).

This departure from traditional understanding of theology, which is based on faith that seeks understanding, is evident in the way Gustavo Gutierrez defines theology. In his view, theology is "critical reflection on historical praxis." Hence theology should engage and address human sociopolitical history and contexts. Theology is a "dynamic, ongoing exercise involving contemporary insights into knowledge (epistemology), man (anthropology), and history (social analysis)."[79]

Liberation theology is based on class struggle since it assumes existence of class structure in a society whereby one class is exploited and marginalized by the other. Consequently, liberation theology adapted Marxist analysis of society, using the language of the bourgeois and proletariat. However, the main objective of liberation theology is transformation of society. Like Marxism, liberation theology tends to "condemn religion for supporting the status quo and legitimating the power of the oppressor. But unlike Marxism, liberation theology turns to the Christian faith as a means for bringing about liberation."

Liberationists claim that they only use Marxism for socioeconomic analysis of the society but they remain loyal to the traditional Christian teaching since "human liberation may begin with the economic infrastructure, but it does not end there."[80] The main premise on which Liberation Theology is based is Marxist, that is, existence of socio-economic and political strata or classes in the society. Like Marxism, liberation theology, though not violently, call for both bourgeois and proletariat to eliminate socio-economic and political strata.

Liberation theology tends to avoid some fundamental facts of spirituality as it emphasizes on Christ's sacramental presence in the poor and oppressed. The plight of the poor and marginalized becomes in a very realistic sense crucifixion of Christ. While liberation theologians teach love of all humans, they claim that their struggle is against inhuman treatment of humans by fellow humans. Such treatment is not only unchristian; it is contradictory to the essence of humanity. "From biblical perspective liberationists argue that the poor represent the totality of the 'Other,' or God. Treatment to the poor is treatment to God since God identifies with the poor and is represented by the poor."

Liberation theology equates biblical salvation with liberation of human beings from oppressive social structures and injustice by other humans. "Liberation theology for all practical purposes, equates loving your neighbor with loving God. The two are not only inseparable but virtually indistinguishable." God is practically in the neighbor and the treatment given to neighbor is given to God. Thus "The history of salvation becomes the salvation of history embracing the entire process of humanization."

According to Liberation theology, therefore, Biblical liberation of Israelites from Egyptian slavery and Jesus' teaching, death and resurrection "stand out as the prototypes for the contemporary human struggle for liberation. These biblical events signify the spiritual significance of secular struggle for liberation."[81] Jesus' teaching

---

[79] Webster (Accessed 12 May 2012).
[80] Webster (Accessed 12 May 2012).
[81] Webster (Accessed 12 May 2012).

and example make present what happens to God when injustice and oppression happens to a human person.

Liberation theology tends to resist both abstraction and objectification of God from creation.[82] "Essentialism is replaced with the notion of Jesus' relational significance." Jesus does not only show us the way to God or the way to become children of God, he reveals to us the way to become human. "The meaning of Jesus' incarnation is found in his total immersion in a historical situation of conflict and oppression." In his immersion into humanity and assumption of human situation, Jesus "absolutizes the values of the kingdom, unconditional love, universal forgiveness, and continual reference to the mystery of the Father."[83] The basic idea of liberation of human beings from oppression is good and central to the Christian message and mission.

Thus liberation theologians and believers have experienced support of some Latin American bishops. This support is evident in the 1968 conference in Medellin, Columbia. However, liberation theology has major problems in its methodology. Adoption of Marxist communist style of society analysis rather than sin and its alienating character is one of the major problems of liberation theology. The second major problem is its implied relativism with regards to Christ's ontological existence and its significance in itself, independent of its analogical and incarnational association with the poor and the marginalized. Within Liberation Theology God is almost not independent of creation. Liberation theology implies that conception of God is conditioned by history and context. Implicitly, "To argue that our conception of God is determined by the historical situation is to agree with radical secularity in absolutizing the temporal process, making it difficult to distinguish between theology and ideology."[84]

Ubuntu shares into the liberation theology's desire to eliminate socio-economic stratification. Ubuntu also believes in God's immanent nature. However, Ubuntu believes also in God's transcendent nature and his separate distinct nature and existence within and without the world. In other words, Ubuntu is much more like the mainstream catholic social teaching.

Being human, both the rich and the poor deal with sin as estrangement or alienation from oneself, from the society and from God. Liberationists, however, over-emphasize the poor making them "not only the object of God's concern but the salvific and revelatory subject. Only the cry of the oppressed is the voice of God. Everything else is projected as a vain attempt to comprehend God by some self-serving means." However, all human beings are children and images of God. God's salvation is for all God's children and God is accessible to all humans who intend to reach and relate with Him. God does not overtake the person and autonomy of the poor, so that the poor ceases to exist.

The poor don't become God in the sense that substantially, they remain not any more or any less human as the rich. "Biblical theology reveals that God is for the poor, but it does not teach that the poor are the actual embodiment of God in today's

---

[82] Webster (Accessed 12 May 2012).
[83] Webster (Accessed 12 May 2012).
[84] Webster (Accessed 12 May 2012).

world." There is an actual danger within brands of liberation theology to humanize God in the poor, which tends to relativize God's ontological reality. That is heretical.[85] Apart from its wrong dogmatic assumptions, liberation theology is pastorally relevant, especially in its fight against exploitation and marginalization of the poor and for inclusion of all human beings as equal members in the mystical body of Christ, the Church.

In sum, liberation theology, by overemphasizing the significance of Jesus' incarnation, its sociological, anthropological and economic significance, especially his analogical identification with the poor, tends to compromise not only Jesus' objective reality as God and man, with a real ontological existence. As the second person of the Holy Trinity, Jesus transcends human history. As God incarnate he is fully God and fully human. No other human being can actually ontologically be the person of Jesus. He remains the ideal of humanity which, though incarnate and immanent, transcends humanity, and which remains a mystery to humans due to his divine-human nature. Some liberation theologians have a problem with the doctrine of the revealed Jesus' deity. Liberation theology claims that Jesus "is different from us by degree, not by kind, and that his cross is the climax of his vicarious identification with suffering mankind rather than a substitutionary death offered on our behalf to turn away the wrath of God and triumph over sin, death, and the devil."[86]

Jesus' historical life and the paschal mystery cannot be separated from his nature as God-man. Jesus' death and resurrection is both for the rich and for the poor. It is for all. Paschal mystery is incomprehensible without its linkage with God's love for all humans and the reality of a sinful humanity in need of redemption. "A theology of the cross which isolates Jesus' death from its particular place in God's design and shuns the disclosure of its revealed meaning, is powerless to bring us to God, hence assuring the perpetuity of our theological abandonment."[87] Liberation theology shares with Ubuntu the perspective of human sacred equality before God. Nobody should be marginalized. By nature a human person is irreducible to a mere means for another human person. Marginalization of a human person depletes the very essence of humanity of meaning. Self-realization of humans, whether moral, cognitive or sociological, is never independent of other human beings.

The essential contingency of humans to fellow humans is the core of Ubuntu worldview. Contrary to the Liberation Theology, Ubuntu neither assumes nor views human community in terms of classes of the rich and the poor. As an ethic, Ubuntu points to the responsibility of the rich towards the poor as a matter of commonsense, justice and morality. One big difference between Ubuntu, liberation theology and the main stream Catholic socio-ethical tradition is that Ubuntu is already unanimously accepted and lived as a way of life. It has been practically learned over many years and has been orally passed on from one generation to another. Thus, Ubuntu is naturally more real and practicable than both liberation theology and the mainstream catholic social teaching.

---

[85] Webster (Accessed 12 May 2012).
[86] Webster (Accessed 12 May 2012).
[87] Webster (Accessed 12 May 2012).

5.2.1.2   John XXIII's *Pacem in Terris*—Authentic Peace Must Be Based on Observance of Human Rights and Justice

The central point and argument of the encyclical *Pacem in Terris* of John XXIII is that peace remains an empty word if is not based on truth, justice in accordance with the requirements of human rights, charity, and human freedom. [88] In other words, there is no peace if there is no justice charity and freedom. Central to the meaning and content of justice is human rights. John XXIII provides a hierarchy of human rights. Basic human rights are "right to life, bodily integrity, food, clothing, shelter, rest, medical care, necessary social services … the right to respect for one's person, good reputation, freedom to search for truth, freedom of speech, freedom of information, the right to share in the benefits of culture and education."[89]

Other important human rights include: freedom of worship; freedom to choose one's state of life and to form a family; freedom of initiative in the economic field; right to work; right to safe and humane working conditions; fair wage; private property; freedom of assembly; freedom of association; freedom of movement and residence; right to legal protection of human rights; and right to participate in human affairs; freedom of worship, and the right to act freely and responsibly.[90] These rights ought to be respected and effectively fulfilled by all human beings if peace is to be achieved.[91]

John XXIII praised the United Nations for its Universal Declaration of Human Rights. In his view, the declaration represents an important step on the path to juridico-political organization of the world community.[92] John XXIII explores the meaning and implication of rights. Rights are reciprocal and societal in nature, in the sense that, they obligate others to acknowledge and respect them. If well observed, therefore, there will not only be peace but also a stronger human society. There is, for example, an obvious social dimension to the right to respect private property which becomes a duty to those who observe it. In sum, rights imply duties and obligations, thus tying humans together for the sake of order and peace. Tied to the concept of duty and obligation that rights imply is individuals' mandate to contribute to the common welfare, which should not be left aside. Actively participating in government for both self-interest and common good is part of the obligations that human rights imply.[93]

---

[88] Pope John XXIII. Encyclical letter on peace on earth *Pacem in Terris*. http://www.vatican.va/holy_father/father/john_paul_ii/encyclicals/documents/hf_j-xxiii_enc_11041963_pacem_en.html, p. 167.

[89] Pope John XXIII. Encyclical letter on Peace on Earth *Pacem in Terris*, pp. 11–13.

[90] Pope John XXIII. Encyclical letter on Peace on Earth *Pacem in Terris*, p. 15, 18–21, 25–26, 32, 37, 96.

[91] Pope John XXIII. Encyclical letter on Peace on Earth *Pacem in Terris*, p. 32.

[92] Pope John XXIII. Encyclical letter on Peace on Earth *Pacem in Terris*, p. 144.

[93] Pope John XXIII. Encyclical letter on Peace on Earth *Pacem in Terris*, p. 22, 28, 30, 53, 73, 146, 150.

Just as human individuals have rights, human communities also have rights that should be observed and respected by individuals and other communities, John XXIII points out that each country has a right to existence, a right to self-development and the means to attain it, a right to a good name and due respect. Equal dignity of all people implies elimination of personal and racial prejudice and discrimination. Human equality means, by implication, racial equality. There should not be any trace of racism in international relations.[94] Inevitably attached to national human rights are the corresponding duties to respect the same rights in other countries.

Moreover countries, have duties to accept immigrants and integrate them as new members of their society.[95] Just as human persons are equal in dignity, nation states are equal in dignity with corresponding and reciprocal rights and duties.[96] Nations' relationships ought to be harmonized in truth, justice and freedom.[97] John XXIII warns that for the sake of global peace, economically or technologically developed countries should not take advantage of their superior position over other nations.[98] Moreover, like personal development, national development should not be based on exploitative or oppressive relationship with other nations.[99]

John XXIII points to the role of public authority of the world community whose fundamental objective is to recognize, respect, safeguard and promotion of human rights of all persons and nations.[100] One of the most important element missing in the global community of nations is exactly what John XXIII alludes to, that is, "public authority of the world community." United Nations can only declare the rights and possibly condemn nations, national leaders and persons who do not comply. It lacks legislative and executive authority to oversee implementation of human rights. This situation renders it toothless in the face of national or personal refusal to comply.

The objectives of John XXIII's *Pacem in Terris* are substantially similar to the objectives of Ubuntu. Human inevitable interrelationships that should be guided by human rights based on human dignity, observed, and implemented by all persons and the government is the ideal of Ubuntu. Although Ubuntu does not explicitly mention the adverse effects of refusal to respect human rights and observe justice, its strong emphasis on respecting human dignity and basing personhood on individual's dealings with other individuals means that human rights are the very kernel of Ubuntu. John XXIII's *Pacem in Terris'* recognition of the importance of global human solidarity and cooperation gives shape to the essence of Ubuntu which emphasizes the human genre's uniqueness but also the importance of the entire species to be united in recognition of each member of the race, that is, every human person. Ubuntu also emphasize the role of nature to the human species and humans' individual and common responsibility to protect and cherish it.

---

[94] Pope John XXIII. Encyclical letter on Peace on Earth *Pacem in Terris*, p. 44, 86.

[95] Pope John XXIII. Encyclical letter on Peace on Earth *Pacem in Terris*, p. 86, 92, 106.

[96] Pope John XXIII. Encyclical letter on Peace on Earth *Pacem in Terris*, p. 86, 89.

[97] Pope John XXIII. Encyclical letter on Peace on Earth *Pacem in Terris*, p. 80.

[98] Pope John XXIII. Encyclical letter on Peace on Earth *Pacem in Terris*, p. 88.

[99] Pope John XXIII. Encyclical letter on Peace on Earth *Pacem in Terris*, p. 92.

[100] Pope John XXIII. Encyclical letter on Peace on Earth *Pacem in Terris*, p. 139.

5.2.1.3   John Paul II on Centrality of Ethics in Development (*Sollicitudo Rei Socialis*)

In *Sollicitudo rei Socialis* John Paul II applies the teachings of Vatican II Council to specific social problems. Specifically, the document deals with the problems of development and underdevelopment at all levels of society but mainly at the level of international socio-economic relations.[101] According to Vatican II council and John Paul II's *Sollicitudo rei Socialis* being social is natural to human beings. However, growth in sociality and sociability is a vocation of all humans which aims at making each human an active player and responsible builder for the earthly society.[102] Sollicitudo rei Socialis reiterated, explicated and reinforced Vatican II's social question which underlines the duty of solidarity between the rich and the poor for the good of both groups.[103]

According to John Paul II, there cannot be authentic human societal development if there is no peace. There cannot be true peace if there is no justice. The more peaceful any particular society is, the more developed it is. In his view, as he explicitly stated, "development is the new name of peace." Hence, arms race is a clear sign of lack of peace, thus lack of authentic human development.[104]

According to John Paul II one of the greatest global injustices is that of "poor distribution of the goods and services originally intended for all." This is a structural societal ethical problem that the pope warns against, especially because it wrongly compromises human essential dignity basing human worth on possessions. Ontologically, "being" is basic and primary while "having" is secondary to "being." "Having" does not contribute to human perfection unless it contributes to maturing and enriching of "being."[105] Moreover, "having" can detract from "being."

This reversal is unethical. Human beings are fundamentally worth more than whatever they possess.[106] Some of the problems which explicate this ethical problem are reflected in the housing crisis partly due to urbanization[107] and growing numbers of unemployment which begs the question on the type of development

---

[101] Pope John Paul II. Encyclical letter on the concern of the church for the social order *Sollicitudo Rei Socialis*. http://www.vatican.va/holy_father/john_paul_ii/encyclicals/documents/hf_jp-ii_enc_30121987_sollicitudo-rei-socialis_en.html, p. 7.

[102] Pope John Paul II. Encyclical letter on the concern of the church for the social order *Sollicitudo Rei Socialis*, p. 1.

[103] Pope John Paul II. Encyclical letter on the concern of the church for the social order *Sollicitudo Rei Socialis*, p. 9.

[104] Pope John Paul II. Encyclical letter on the concern of the church for the social order *Sollicitudo Rei Socialis*, p. 10.

[105] Pope John Paul II. Encyclical letter on the concern of the church for the social order *Sollicitudo Rei Socialis*, p. 28.

[106] Pope John Paul II. Encyclical letter on the concern of the church for the social order *Sollicitudo Rei Socialis*, p. 28.

[107] Pope John Paul II. Encyclical letter on the concern of the church for the social order *Sollicitudo Rei Socialis*, p. 17.

being pursued.[108] Denying human interdependence is unrealistic. Human interdependence cannot be separated from ethical requirements for the rich and the poor of any human community and between rich and poor countries.[109]

John Paul II points out some of the structural problems that account for unethical human economic inequality and its escalation. One of the problems mentioned is that of omissions on the part of developing countries. This problem is doubled by lack of ethical response by affluent world and manipulation of economic, political and social mechanisms to benefit the already wealthy at the expense of the poor.[110] John Paul II regrets that the human right of economic initiative in service of individual and common good is too often suppressed, which in turn, frustrates people's creativity.

Such kind of totalitarianism reduces people into a mere means for other people, that is, making them "objects." Reduction of human persons into objects by, and for other persons is one of the greatest immoralities against human nature. There are other forms of immoralities against the very nature of humanity such as denial of human right to worship and racial discrimination. Such immoralities contradict human social and ethical nature.[111]

John Paul II warns against international global structural injustice which should be addressed. One of such injustice is the intentional ignorance of the developed world to the abject poverty in the developing world. John Paul II observes that production and distribution of global goods is not fair. The developing world is falling behind the developed world and unity of the human species is compromised by divisions into first, second and third world countries.[112] There is a vicious cycle of poverty on the international level in which debtor nations are increasingly forced to export capital. This situation aggravates underdevelopment.[113] Another equally threatening situation is the division of the world into liberal capitalism of the west against Marxist collectivism of the East.[114] Exaggeration of concern for national or regional security may be a stumbling block to human cooperation, thus undermining human rights.[115]

International trade system, notes John Paul II, discriminates against the poor countries and international division of labor exploits workers in the interest of profit

---

[108] Pope John Paul II. Encyclical letter on the concern of the church for the social order *Sollicitudo Rei Socialis*, p. 18.

[109] Pope John Paul II. Encyclical letter on the concern of the church for the social order *Sollicitudo Rei Socialis*, p. 17.

[110] Pope John Paul II. Encyclical letter on the concern of the church for the social order *Sollicitudo Rei Socialis*, p. 16.

[111] Pope John Paul II. Encyclical letter on the concern of the church for the social order *Sollicitudo Rei Socialis*, p. 15.

[112] Pope John Paul II. Encyclical letter on the concern of the church for the social order *Sollicitudo Rei Socialis*, p. 14.

[113] Pope John Paul II. Encyclical letter on the concern of the church for the social order *Sollicitudo Rei Socialis*, p. 19.

[114] Pope John Paul II. Encyclical letter on the concern of the church for the social order *Sollicitudo Rei Socialis*, p. 20.

[115] Pope John Paul II. Encyclical letter on the concern of the church for the social order *Sollicitudo Rei Socialis*, p. 22.

maximization. Technology transfer is also unfair to the poor countries. International organizations need reform, so that they are free from manipulation by political rivalries.[116] Solidarity among developing countries will call for greater cooperation and establishment of effective regional organizations. Regional organizations may help the developing countries defend against exploitation by the developed countries. [117] Omission is a great ethical problem among the developing world. Developing countries must take up their own responsibilities.

They should promote self-affirmation of their own citizens through programs of literacy and basic education.[118] In sum, the Pope warns against human succumbing to struggle for survival and survival of the fittest state of affairs like animals. Human beings are essentially ethical beings. Their moral nature should be reflected in their treatment of one another as individuals, societies or nations.

In spite of the structural injustices in the contemporary world, the pope notes with optimism some signs of hope. Some of the encouraging signs of development include global ecological consciousness and concern; some third world countries have achieved some self-sufficiency in basic human needs such as food; there is global growing awareness of human rights.[119] There is an intimate connection between liberation and development, overcoming obstacles to a "more human life."[120] In sum John Paul II in his *Sollicitudo rei Socialis* systematically articulates the fundamental requirements of ethical societal living based on the assumption and premise of human essentially social nature. In his analysis he points out major global deficiencies which contradict human nature while pointing the way forward in line with Vatican II's recommendations.

John Paul II's concern is in agreement with, and verbalizes Ubuntu perspective of moral human societal living. All in all both John Paul II and Ubuntu agree on the need to cherish human solidarity which is unachievable without justice and peace. Justice and peace should give special preference to the incapacitated members of the human race and the poor, based on their undeniable essential dignity as humans. Substantially similar between John Paul II's encyclical and the philosophy of Ubuntu is the philosophical and realistic distinction between being and having. Both John Paul II and Ubuntu philosophy underline the importance of being over having and that having should be at the service of being. If this distinction is globally acknowledged there could be a reduction in the problem of marginalization and increasing exploitation of the poor by the rich. According to Ubuntu possession of property is supposed to be for use by the one possessing but should be given out to

---

[116] Pope John Paul II. Encyclical letter on the concern of the church for the social order *Sollicitudo Rei Socialis*, p. 43.

[117] Pope John Paul II. Encyclical letter on the concern of the church for the social order *Sollicitudo Rei Socialis*, p. 45.

[118] Pope John Paul II. Encyclical letter on the concern of the church for the social order *Sollicitudo Rei Socialis*, p. 44.

[119] Pope John Paul II. Encyclical letter on the concern of the church for the social order *Sollicitudo Rei Socialis*, p. 26.

[120] Pope John Paul II. Encyclical letter on the concern of the church for the social order *Sollicitudo Rei Socialis*, p. 46.

help the one who does not have. Recognition of human equality of being is important in both philosophies.

5.2.2   *Mutuality*

Another equally essential concept of social cohesion is that individual and common societal realization requires human mutuality.[121] Human fulfillment can only
happen in society since only in society is a "firm and persevering determination to
commit oneself to the common good" possible.[122] Thus, human actions should be
characterized by a mutuality of concern for oneself and others.[123] A genuine sense of
development interprets the individual within a mutual framework of cooperation in
which "self-fulfillment is not juxtaposed with the fulfillment of others." The inevitable interdependence of humans necessitates commitment to collaboration.[124] Mutuality does not overlook individual needs, rather it places individual needs within
an essential relational context that views "fulfillment of self in and through the
flourishing of other."

Authentic personal fulfillment happens in collaboration and cooperation with
others. Personal fulfillment contributes to the good of the society as a whole.[125] Leo
XIII, in his encyclical *Rerum Novarum,* explicates what genuine mutuality means.
It means acknowledgement of equality and equity, assurance of just returns from
human labor, decent minimum for all and assurance of government security and
protection for the less privileged. He argues that it is in the interest of the wealthy,
the poor and the government to ascertain such mutuality.[126] Chapter 2 of this work
shows why mutuality is within the kernel of Ubuntu. Church's tradition on mutuality is inspired by the great commandment, loving one's neighbor as oneself.

Mutuality is recognizing equality and overcoming our natural inclination to ego-
centricism. In both Catholic tradition and Ubuntu culture human action has dual
characteristics. It is at once for self and for the society. Catholic tradition is inspired
by Christ's giving of himself for human community; Ubuntu is inspired by the fact
that all that a human person claims to be his he has received. Giving back is a matter
of justice. Human life is about constant giving and receiving.

---

[121] Pope John Paul II, Encyclical letter on the concern of the church for the social order *Sollicitudo
Rei Socialis*, p. 38.

[122] Pope John Paul II, Encyclical letter on the concern of the church for the social order *Sollicitudo
Rei Socialis*, p. 38.

[123] *Catechism of the Catholic Church*, 1939–1942.

[124] Dorr (1989, pp. 143–154).

[125] Pope John XXIII. Encyclical letter on mother and teacher *Mater et Magistra*, p. 65, *Catechism
of the Catholic Church*, p. 1604, 1880, 1905–1908.

[126] Pope Leo XIII. Encyclical letter on the new things *Rerum Novarum*, p. 51.

5.2.2.1   Pontifical Council for Justice and Peace on Mutuality as Human Nature and Responsibility

The Pontifical Council for Justice and Peace underlines communication as an important aspect of global solidarity. The council condemns exclusion of some parts of human populations from some important communication noting that "communication structures and policies, and the distribution of technology are factors that help to make some people 'information rich' and others 'information poor' at a time when prosperity, and even survival depends on information."

There should not be intentional discrimination in the supply and consumption of useful information. All peoples do not only need to be informed but also need to be trained in using modern communication technologies.[127] Information, participation and inclusion of all people should be the objective of communication. Human beings being by nature societal, mutuality, which is enhanced by effective communication, is an ethical duty of the media. Media's audiences should also play their part in the use of the information communicated both for personal and for common good.

The pontifical council for justice and peace urges political leaders and economists to rethink the urgency of addressing the fact that the "present economic, social and cultural structures are ill-equipped to meet the demands of genuine development." Political participation and social justice are necessary for lasting peace. Solidarity "must be made an integral part of the networks of economic, political and social interdependence that the current process of globalization tends to consolidate."[128] There is no alternative to creating harmonious, peaceful and functional society without elements that facilitate solidarity.

On the level of individuals, each person's need to participate by pursuit of both individual good and common good. Christians who have been elected and trusted with public offices should appreciate and promote democracy, that is, assuring "participation of citizens in making political choices." Ascertaining assumption of, and removal from, office in accordance to the will of the people who are free in conscience and responsible for their government is an important sign of political maturity and solidarity.[129]

According to the Pontifical Council for justice and peace morality is an "absolute necessity" for political or public service. Ignoring the moral dimension of public service "leads to dehumanization of life in society and of social and political institutions." Every person needs to live and act in accordance with his conscience. If one fails to follow the dictates of his conscience or contradicts his conscience, he can achieve neither happiness nor authentic fulfillment.[130]

However, morality is not only a matter of personal conscience, it is also social. Actually, every society has some sort of moral principles. Complete absence of moral principles annihilates society. The most important question is whether the

---

[127] Pontifical Council for Justice and Peace (Accessed 15 April 2012) no. 561.

[128] Pontifical Council for Justice and Peace (Accessed 15 April 2012) no. 564.

[129] Pontifical Council for Justice and Peace (Accessed 15 April 2012) no. 567.

[130] Pontifical Council for Justice and Peace (Accessed 15 April 2012) no. 566.

principles in place or imagined by a people or their leaders is objectively ethical. The *Catechism* teaches that humanity is inherently social and that the individual person by nature requires a community of other persons to fulfill his or her person-hood.[131]

The best way to understand the message of the Pontifical Council for Justice and peace is provided by the Catechism of the Catholic Church. The catechism teaches that natural nuclear human family is the foundation and extension of the larger fam-ily of the entire race.[132] Thus, all human beings, the members of the larger human family should be perceived as "neighbors" who have "inherent rights and mutual responsibilities for the wellbeing of the society."[133] Any human being is at once personal and social. Persons are simultaneously individual and social, that is unique beings created within a social context.[134]

Ubuntu, the Pontifical Council for Justice and Peace acknowledges the relation-ality of all humans, interdependence of all humans, mutual need of all humans for the cosmos, necessity of morality for social life, and the need to respect individual conscience in the context of societal responsibility. Ubuntu realizes that it is through communication that human beings become a means of both their own self-actual-ization and that of others. Communication helps the on-going realization of the fact that all human beings are equal and that they not only need to share the resources of the planet earth, they need to share each other as they need each other.

5.2.2.2 Benedict XVI's *Caritas Veritate*—on Mutuality

The general theme of Benedict XVI's *Caritas Veritate* is the undeniable fact of the centrality of truth in Christian faith. Even though the main message of Christianity is charity, charity has to be founded on truth. God is love and God is truth. God's Love is the truth without which nothing makes sense. Benedict XVI analogically equates "A Christianity of charity without truth would be more or less interchangeable with a pool of good sentiments, helpful for social cohesion, but of little relevance."[135] A genuine sense of truth is necessary for authentic development. Without truth "the social action ends up serving private interests and the logic of power, resulting in social fragmentation."[136] Even though the church "does not have technical solutions to offer," she does have "a mission of truth to accomplish" for, and about human society which "is attuned to man, to his dignity, to his vocation."[137]

---

[131] *Catechism of the Catholic Church*, 1878–1880.

[132] *Catechism of the Catholic Church*, 1882.

[133] *Catechism of the Catholic Church*, 1889.

[134] Catechism of the Catholic Church, 1879.

[135] Pope Benedict XVI. Encyclical letter on charity in truth *Caritas in Veritate*. http://www. vatican.va/holy_father/benedict_xvi/encyclicals/documents/hf_ben-xvi_enc_20090629_caritas-in-veritate_en.html, pp. 1–4.

[136] Pope Benedict XVI (2009, p. 5).

[137] Pope Benedict XVI (2009, p. 8–9).

Benedict XVI then turns his attention to the modern world's tendency to misinterpret true human development, especially by equating it with mere material well-being, usually at the expense of other human beings' subsistence. As such, writes Benedict XVI, development "has many overlapping layers." Genuine development has to be holistic and integrative of the entire human family. This is not often the factual situation. In the words of Benedict XVI, "the world's wealth is growing in absolute terms, but inequalities are on the increase." There are new forms of poverty and marginalization of majority of human beings by what is interpreted as development of a minority of the human race. The Pope laments that there is an increase in corruption both in rich and poor countries.

There is still the almost chronic problem of disrespect for the rights of workers. Benedict notes that even international aid "has often been diverted from its proper ends, through irresponsible actions." There is an increasing problem of monopoly of knowledge, an "is excessive zeal for protecting knowledge on the part of rich countries, through an unduly rigid assertion of the right to intellectual property, especially in the field of health care."[138] There is a growing problem of "unregulated exploitation of the earth's resources" without enough consideration of its effect on both current and future generations. This crisis calls for what Benedict XVI calls "humanistic synthesis." The crisis glaring at our present generation, using the words of Benedict XVI, "obliges us to re-plan our journey."[139]

Addressing the scandal of hunger in the world, Benedict XVI calls for a "network of economic institutions" to work together for resolution of this problem which is a scandal. He believes that the solution lies in "employment of relevant and effective techniques of agriculture and land reform in the third world countries."[140] Human dignity should always be protected from the shame of hunger and starvation. Benedict reminds the world that economics is not free from principles of ethics and morality. If development "is to be authentically human," it has "to make room for the principle of gratuitousness," especially with regards to market principles of supply, demand and profit maximization.[141] Divorcing morality and principles of ethics from economics leads to suppression of human life.

However, "When a society moves towards the denial or suppression of life, it ends up no longer finding the necessary motivation and energy to strive for man's true good."[142] Realistic human development "depends above all on a recognition that the human race is a single family."[143] The market, civil society, and the state should work together to ascertain human solidarity and dignity.[144] Government should respect labor unions, especially because they safeguard human dignity. Benedict XVI contends, "The primary capital to be safeguarded and valued is man,

---

[138] Pope Benedict XVI (2009, p. 22).
[139] Pope Benedict XVI (2009, p. 21).
[140] Pope Benedict XVI (2009, p. 27).
[141] Pope Benedict XVI (2009, p. 34).
[142] Pope Benedict XVI (2009, p. 28).
[143] Pope Benedict XVI (2009, p. 53–56).
[144] Pope Benedict XVI (2009, p. 35–39).

the human person in his or her integrity."[145] Dealing with human persons directly demands utmost care. Benedict urges nations to work together in addressing the problem of migration because of the human dignity of the immigrants. Immigrants, like other citizens possess "fundamental, inalienable rights that must be respected by everyone and in every circumstance."[146]

Benedict XVI mentions the importance of bioethics in safeguarding human dignity. In his view, bioethics is of crucial importance as it functions as a battleground between the "supremacy of technology and human moral responsibility." However, bioethics has to be inspired and motivated by faith since "Reason without faith is doomed to flounder in an illusion of its own omnipotence." The Pope then exhorts the human race to have "new heart" in order to rise "above a materialistic vision of human events."[147]

Once again, Benedict XVI's views analyzed in *Caritas in Veritate* reflect the shared perspective between Ubuntu and Catholic Social Teaching concerning human dignity; necessity of safeguarding the truth about human dignity; ethical precedence of ethics and morals over materialism and a worldview of connectedness of the human species as one family whose familial bonds have to be preserved and protected as a matter of truth and ethics.

As the Bible categorically states, "truth shall set you free," (John 8:32) there cannot be true freedom, which is not founded on truth. Real human freedom must recognize the truth of the humanity and humanness of other persons along with the rights that come with that recognition, hence the obligation due to them. All this teaching is contained in the maxims, *I am because you are*; *I am because we are*; *a human being is human because of the otherness of other human beings*. This ontological, psychological, sociological and epistemological truth leads to much truth about human beings which should help them be even freer in their relationship with and treatment of other. It all boils down to moral treatment of other persons.

5.2.2.3 United States Bishops on Cooperation and Partnership for the Public Good

The Catholic bishops of the United States of America, in their pastoral letter on economic justice remind their nation that economic prosperity is a product of human beings. It results from "the labor of human hands and minds." Economy being a product of human beings, it cannot be divorced from ethical principles that characterize and guide human activity. The economy "is not a machine that operates according to its own inexorable laws, and persons are not mere objects tossed about by economic forces."

There is special relationship between human persons and the work they do. While the product of the work they do may help to sustain and improve human life, human

---

[145] Pope Benedict XVI (2009, p. 23–25).

[146] Pope Benedict XVI (2009, p. 62–64).

[147] Pope Benedict XVI (2009, p. 74–77).

beings achieve self-realization through their work. "All work has a threefold moral significance. First, it is a principal way that people exercise the distinctive human capacity for self-expression and self-realization. Second, it is the ordinary way for human beings to fulfill their material needs. Finally, work enables people to contribute to the well-being of the larger community." Human labor is social in the sense that it is not merely for oneself. It has a social dimension that benefits others.[148] Human beings ought to work since they need to participate and contribute into the common good and social justice. Human labor though, should be justly rewarded.[149]

The United States' bishops interpreted and applied the teaching of the Second Vatican Council for the people of their country about value and role of work to individuals and society. In *Gaudium et Spes*, for example, work is explained as a matter, not only of free choice but as a matter of justice. The document states, "The best way to fulfill one's obligations of justice and love is to contribute to the common good according to one's means and the needs of others."[150] Work, therefore is an obligation to all that can work. Those who have the ability to work should support those who cannot work because of some physical or mental impediment. The United States bishops applied the ethical Principle of Subsidiarity to explain both the necessity of work for all human beings and the ideal of fairness to the concept and substance of humanity that all humans share. Recognition, acknowledgment and participation are of utmost importance since they relate to the meaning of human life. However, people must participate in contributing to the common good according to their ability.

In their pastoral letter, the United States bishops re-presented the principle of subsidiarity with the intention of explaining justice in participation to the common good. The principle of subsidiarity as formulated by the bishops states, "Just as it is gravely wrong to take from individuals what they can accomplish by their own initiative and industry and give it to the community, so also it is an injustice, and at the same time a grave evil and disturbance of right order, to assign to a greater and higher association what lesser and subordinate organizations can do."[151]

The implication of the principle is that social activity ought to help (*subsidium*) both its subject and all members of society rather than absorb or marginalize any of them. If the principle of subsidiarity is applied well there will be harmony in the society for a number of reasons. The principle ascertains freedom and recognition of societal members by minimizing class struggle; it favors institutional and societal pluralism, encourages initiative and creativity and increases solidarity.

The principle of subsidiarity recognizes everybody and every institution as a significant part of the functioning of the society. In and of itself, recognizing the worth

---

[148] United States Bishops, "Economic justice for all: Pastoral letter on catholic social teaching and the U.S. economy", p. 97.

[149] United States Bishops, "Economic justice for all: Pastoral letter on catholic social teaching and the U.S. economy". http://www.usccb.org/upload/economic_justice_for_all.pdf, no. 96.

[150] Second Vatican Council (1965, p. 30).

[151] United States Bishops. "Economic justice for all: Pastoral letter on catholic social teaching and the U.S. economy", p. 99.

of each person or institution helps promote peace and justice, leave alone enhancing the economy and social living.[152] The ideal of a good society that the Principle of Subsidiarity targets is that envisioned by Vatican II Council's *Gaudium et Spes*, "the entire *human family* seen in its total environment ... the world as the theatre of human history." Whoever is human is a unique member of the human family to be recognized and cherished.[153] Ubuntu worldview, is an application of the principle of subsidiarity, not only because of its societal division of labor but, especially, in the recognition of the contribution of human *otherness* to every human person. According to the ideal of Ubuntu no one has nothing to offer; everybody should be keen to recognize what everybody else offers.

5.3 Minority Empowerment

Another significant component of Catholic ethical tradition is based on minority empowerment. This component flows right from the central teaching of Jesus Christ who identified with the minority and emphasized equality of people and the importance of service to the minority. He paradoxically stated that "the last shall be first and the first shall be last" (Mt. 20:16). This component consists of two important concepts. First, the principle of subsidiarity, that each member of society has a right to be helped to participate in the common good according to his potential and ability; second, protection of the vulnerable and recognition of right to healthcare for all.

5.3.1 Subsidiarity

The first concept of minority empowerment is the principle of subsidiarity. As it has been demonstrated in the previous section the Catholic Church advocates for creation of an environment whereby each member of society is helped to participate in the common good according to his potential and ability.[154] The Church condemns "possessive individualism and freedom of indifference".[155] Respect for the common good and human dignity requires the practice of the principle of subsidiarity. The principle of subsidiarity received its classic formulation from Pope Pius XI in his 1931 encyclical, *Quadragesimo Anno*. The principle aims at the recognition and utilization of the potential of each component of society.

---

[152] United States Bishops, "Economic justice for all: Pastoral letter on catholic social teaching and the U.S. economy", p. 100.

[153] Gaudium et Spes (1965, p. 2).

[154] Pope Pius XI. Encyclical letter on in the fortieth year *Quadragesimo Anno*, John Paul II. 1991. Encyclical letter on in the hundredth year *Centesimus Annus*. http://www.vatican.va/holy_father/john_paul_ii/encyclicals/documents/hf_jp-ii_enc_01051991_centesimus-annus_en.html, p. 13.

[155] Werpahowski (1984, pp. 81–115).

The principle of subsidiarity protects self-realization of individuals and subordinate organizations.[156] In *Centesimus Annus,* John Paul II explained the social teaching of the church in accordance with the principle of subsidiarity as stated in *Rerum Novarum.* The pope explained that the social nature of the human person is not completely fulfilled in the state, but is realized in various intermediary groups, beginning with the family and including economic, social, political and cultural groups which stem from human nature itself and have their own autonomy, always with a view to the common good.[157]

In their 1986 pastoral letter on the economy and poverty, the United States Catholic Bishops articulated the principle of subsidiarity in their terms. The government should help smaller organizations and individuals "contribute more effectively to social well-being and supplement their activity when the demands of justice exceed their capabilities.[158] The United States bishops emphasized that the government should be proactive and take an enabling initiative. This aspect of the principle makes an active rather than passive principle.[159] Consequently, common good cannot be fully realized without implementing the principle of subsidiarity. The culture of Ubuntu is founded on subsidiarity, not only between human beings and human organizations, but also within nature itself.

Each potential should be helped into realization. The principle of subsidiarity protects the poor and lower class populations in a very special way by enabling them to realize their potential. It gives every organization and person a chance to participate in the common good. Consequently it is important for solidarity and fostering of human dignity. The principle of subsidiarity is at the core of Ubuntu. It is reflected in the Sub-Saharan common division of labor according to age, gender, talent, physical strength and disability. In Ubuntu the principle aims at affirming and empowering everybody so that each person is recognized for his contribution and participation in the flourishing of the society.

5.3.1.1 Pius XI's *Quadragesimo Anno*—Extreme Forms of Both Socialism and Capitalism are Auto-destructive

One of the dangers of exaggerated capitalism is its tendency to be auto-destructive so that the whole economic and social system crushes and collapses. Pius XI in his *Quadragesimo Anno* warns that capital had, and still were appropriating too much to itself. Economic institutions were moving in the direction of giving all the wealth to the rich.[160] Unhealthy disparity between the very few extremely rich and the big majority extremely poor was increasing in such a way that economic life

---

[156] Pope Pius XI. Encyclical letter on in the fortieth year *Quadragesimo Anno*, p. 79.

[157] Pope John Paul II. Encyclical letter on in the hundredth year *Centesimus Annus*, p. 13.

[158] U. S. Catholic Bishops. 1986. *Economic justice for all.* http://www.osjspm.org/economic_justice_for_all.aspx. p. 124.

[159] U. S. Catholic Bishops. *Economic justice for all*, p. 124.

[160] Pope Pius XI. 1931. http://www.vatican.va/holy_father/pius_xi/encyclicals/documents/hf_p-xi_enc_19310515_quadragesimo-anno_en.html, no. 54.

was becoming rapidly inhuman and unethical. Monopolies and economic dictator-ship had taken over the market. Power was becoming increasingly concentrated in the hands of the few rich minorities.[161] According to the Pope, this state of affairs is a fertile ground for conflict, protest and riots on the local, national and interna-tional levels.[162] The system is in itself violent. The majority poor already experience the violence even if it had not erupted into action in most cases. The cause of the violence is the injustice inherent in the division of the society into two opposing classes.[163]

Equally self-destructive is extreme forms of socialism. Pius XI argues that one section of socialism had already degenerated into communism. Communism is espe-cially dangerous because of its inhumanity. Due to its extermination of private owner-ship, which is not natural to the human race, communism bears within itself seeds of self-destruction. One symptom of this self-destructive tendency of communism is the obvious class struggle always going on within it. The Church condemns all forms of communism as incompatible with the Gospel.[164] Although Pius XI does not condemn capitalism as incompatible with the Gospel as he did communism, especially because it is not in itself vicious, its extreme forms can be dangerous to the human race.[165] The system can condone and enable cruel and inhuman structures.[166] Capitalistic economy is thus not immune to exploitation and marginalization of human beings.

Capitalism becomes dangerous when it is completely, or largely under the con-trol of capital and market forces, which operate on the maxim of struggle for sur-vival and survival of the fittest. Profit maximization motive tends to undermine human demands of morality.[167] Unchecked capitalism can take over politics and governments so that the government comes under the control and greed of the few wealthy individuals. This situation can extend itself into a form of governmental international imperialism.[168]

On the local governmental level, extreme forms of capitalism that has been taken over by capital become ruthless in exploiting, marginalizing and ostracizing the poor. Social life of the rich, which had once been highly developed through many different associations, collapse into a regrettable situation in which there is only the government relating to a few individuals.[169] Social order on which right reason is partly based tends to decline and eventually be exterminated when the society gives in completely to market forces. This extermination of social order is fatal to society as such. The negative force behind this disastrous situation is human greed and selfishness.[170]

---

[161] Pope Pius XI. Encyclical letter on in the fortieth year *Quadragesimo Anno*, p. 105, 107, 109.

[162] Pope Pius XI. Encyclical letter on in the fortieth year *Quadragesimo Anno*, p. 108.

[163] Pope Pius XI. Encyclical letter on in the fortieth year *Quadragesimo Anno*, p. 82.

[164] Pope Pius XI. Encyclical letter on in the fortieth year *Quadragesimo Anno*, p. 112, 128.

[165] Pope Pius XI. Encyclical letter on in the fortieth year *Quadragesimo Anno*, p. 101.

[166] Pope Pius XI. Encyclical letter on in the fortieth year *Quadragesimo Anno*, p. 128.

[167] Pope Pius XI. Encyclical letter on in the fortieth year *Quadragesimo Anno*, p. 132.

[168] Pope Pius XI. Encyclical letter on in the fortieth year *Quadragesimo Anno*, p. 109, 132.

[169] Pope Pius XI. Encyclical letter on in the fortieth year *Quadragesimo Anno*, p. 78.

[170] Pope Pius XI. Encyclical letter on in the fortieth year *Quadragesimo Anno*, p. 97.

Morally rightful approach to ownership of property ought to avoid two equally harmful extremes, namely, individualism, which implies minimization or eradication of social character of the right to own property; and collectivism, which means complete rejection or minimization of the right to private ownership of property.[171] Both extremes harm social life by causing harm both to individuals and to the society. Just as private ownership should not be at the expense of the human naturally social characteristic, so is socializing all individual ownership of property is hazardous both to the individual and to the society. While commutative justice demands respect for the property of other individuals, owners of property are obliged in their human conscience, though not by justice, to use their property in a right way.[172] It is unethical, for example, for a businessperson to completely disregard the conditions of his workers by acting purely on self-interest and profit maximization. Profit maximization and self-interest should not be at the expense of human dignity of any person. Moreover, the overall good of the society ought to be respected.[173] In brief, Christian social principles regarding capital and labor must be put into practice.[174]

One of the most important points that Pius XI makes concerns the value of human labor. Without exertion of human labor capital cannot appropriate and grow. Being essential to the production process, human labor should be fairly rewarded. In other words when one person provides the capital and another person provides labor; the fruit of that labor belongs to both the owner of capital and the laborer. Ethically, they should both share the benefits without either of them exploiting the other.[175] The wealthy should invest in such a way that they provide for employment, thus increasing the chances for many to participate in the common good in line with the principle of subsidiarity.[176] Intentional omission of positions within production process that could be occupied by masses of the poor in the society is immoral.

Eliminating or replacing human labor with machines intentionally to deny employment to laborers rather than increase efficiency and inclusion is equally immoral.[177] To avoid immorality due to exploitation, wages paid to workers must be sufficient to support the workers and their families. It must consider the conditions of work, it must not be so exhausting that it is detrimental to the health of the worker, provide for a some surplus which will allow the worker to raise his quality of living, rather than increase dependence, poverty and human misery.[178]

Pius XI does not discourage free competition in the process of production, rather he warns of the dangers of letting free competition determine the fate of human participation in the production process. He brings to attention the social nature of production and the necessity of human participation, not only in his rationality and

[171] Pope Pius XI. Encyclical letter on in the fortieth year *Quadragesimo Anno*, p. 46.

[172] Pope Pius XI. Encyclical letter on in the fortieth year *Quadragesimo Anno*, p. 47.

[173] Pope Pius XI. Encyclical letter on in the fortieth year *Quadragesimo Anno*, p. 101.

[174] Pope Pius XI. Encyclical letter on in the fortieth year *Quadragesimo Anno*, p. 110.

[175] Pope Pius XI. Encyclical letter on in the fortieth year *Quadragesimo Anno*, p. 53, 57.

[176] Pope Pius XI. Encyclical letter on in the fortieth year *Quadragesimo Anno*, p. 50, 51, 57, 74.

[177] Pope Pius XI. Encyclical letter on in the fortieth year *Quadragesimo Anno*, p. 79.

[178] Pope Pius XI. Encyclical letter on in the fortieth year *Quadragesimo Anno*, p. 74.

creativity, but also in his moral and social nature. To ascertain fairness in free competition, especially, in order to protect the weak poor, the Government needs to control and regulate competition in a way that protects the poor.[179] Moreover, protecting the rights of the weak and poor laborers, the government should not take away the right to ownership. The government must fairly determine what can be privately owned and what should not be owned privately.[180]

In sum, Pius XI systematically, though without knowing it elaborates the ideals of Ubuntu worldview especially with regards to justice in production and distribution of property, human dignity that should not be compromised, protection of the poor, common good and moral social cohesion of all persons for the good of all. Interestingly, the indigenous centuries—old Ubuntu philosophy regulates itself against the extremes of both capitalism and communism by discerning what truly matters, that is, human beings and their fulfillment through other human beings. As it has been revealed in the first two chapters of this book, nobody within Ubuntu society would be left to fall below the minimum accepted line of poverty. The following section shows how Paul VI warns the society of the dangers of economic disparity and desire for profit maximization that undermines human dignity.

5.3.1.2 Paul VI on the Development of Peoples (*Populorum Progressio*)

Paul VI's Major Concern in his encyclical *Populorum Progressio* is the ever widening gap between the rich and the poor. The encyclical notices and thus is concerned that the wider the gap between the rich and the poor, the greater exploitation of the poor and the more structural injustice in the system as a whole. Economic inequalities are infectious in the sense that they cause political inequality so that the richer are also the more politically powerful. In the process the poor are constantly being reduced into a position of insignificance by being overly ostracized and depleted of their worth.[181] The harsh modern economics favors the rich by helping to widen the difference in the sense that their favored economic position makes it easier for the rich to enjoy easier and more rapid growth while the poor have to struggle and develop slowly.[182]

The gap between the rich and the poor tends to have opposite effects on the rich and the poor respectively. While it continually empowers the rich to become even more powerful, it escalates the misery of the poor, class tension and conflicts, thus setting grounds for revolts and riots.[183] Promotion of development should guard against widening of the gap between the rich and the poor because doing so is the

[179] Pope Pius XI. Encyclical letter on in the fortieth year *Quadragesimo Anno*, p. 110, 179–80.

[180] Pope Pius XI. Encyclical letter on in the fortieth year *Quadragesimo Anno*, p. 49, 80.

[181] Pope Paul VI. Encyclical letter on the development of peoples *Populorum Progressio*. http://www.vatican.va/holy_father/paul_vi/encyclicals/documents/hf_p-vi_enc_26031967_populorum_en.html, p. 9.

[182] Pope Paul VI. Encyclical letter on the development of peoples *Populorum Progressio*, p. 8.

[183] Pope Paul VI. Encyclical letter on the development of peoples *Populorum Progressio*, p. 29.

opposite of true human progress that development should seek.[184] Since human beings are socially related by their very essence and nature, marginalization of either an individual or a group of human beings has adverse effects on the worth of the race as a whole.

Paul VI explains the meaning of real authentic development as holistic by nature. It has to be integral promoting the good of every person as a member of the human family. In other words, development of some which happens at the expense of others contradicts the very meaning of development, due to the connectedness of the human family, the neediness of community, common good and humans' moral nature. No one can be completely free from the plight of another.[185] Realistic authentic development is always a transitional movement from less human to more human conditions.

If the transition is not leading to more human conditions for an individual and for the society at large, it is not development.[186] Thus, economic development, which is not human development, is not authentic. In fact it is not development at all.[187] The church should, by its teaching and deeds, give priority to the hungry, the miserable, the diseased and the ignorant, that is, all those who are marginalized and ostracized so that they can share in the benefits of civilization for their individual good and the good of the human genre.[188]

On the level of international economic relations, Paul VI notes an unhealthy situation in which rich nations decide the terms which favor them so that rich nations remain rich and poor nations are condemned to perpetual poverty. This condition is evident in international trade.[189] The Pope warns against neo-colonialism. He observes that even though colonialism should have been abolished, there is a worse kind of colonialism which takes advantage of poor nations by making them more dependent thus exploiting them. The Pope states that this kind of neo-colonialism is in the form of political and economic pressures whose ultimate aim is complete control.[190] Paul VI calls for creation of a more human world, that is, a world in which everyone can live a fully human life. Such development demands human solidarity which in turn demands for ethical maturity. Real moral maturity implies human responsibility and obligation towards everyone, even future generations.[191]

Paul VII worldview in the *Populorum Progressio*, shares the same perspective, insight and objective with Ubuntu. One becomes human in solidarity with other humans; hence one needs other human beings to realize his humanity. One needs to care about other humans because caring is the distinguishing human characteristic

---

[184] Pope Paul VI. Encyclical letter on the development of peoples *Populorum Progressio*, p. 33.

[185] Pope Paul VI. Encyclical letter on the development of peoples *Populorum Progressio*, p. 14.

[186] Pope Paul VI. Encyclical letter on the development of peoples *Populorum Progressio*, p. 20.

[187] Pope Paul VI. Encyclical letter on the development of peoples *Populorum Progressio*, p. 73.

[188] Pope Paul VI. Encyclical letter on the development of peoples *Populorum Progressio*, p. 1.

[189] Pope Paul VI. Encyclical letter on the development of peoples *Populorum Progressio*, p. 57.

[190] Pope Paul VI. Encyclical letter on the development of peoples *Populorum Progressio*, p. 52.

[191] Pope Paul VI. Encyclical letter on the development of peoples *Populorum Progressio*, p. 17, 43, 47, 54.

that cannot be put aside. Empathizing with, relating with, and caring about other humans defines one's personhood. In other words, exploitation of human beings by fellow human beings, which has now become an international phenomenon is dehumanizing, thus unethical and immoral.

5.3.1.3 U.S. Catholic Bishops' Catholic Framework for Economic Life

The main thesis of the United States Catholic bishops' framework for economic life is curbing the growing economic injustice, poverty and growing income gaps in their country and by extension in the entire world. The bishops note that the economic injustice crisis is not limited to the United States of America. It is a global issue. The bishops provided an "ethical framework for economic life as principles for reflection, criteria for judgment and directions for action."[192]

The first directive for all Catholics is one of persons' precedence over the economy rather than the other way round. "The economy exists for the person, not the person for the economy." In other words the economy should always be understood as a means to personhood so it should not take over and enslave any person. The importance of this directive is explicated in Vatican II's *Gaudium et Spes*. To underline the precedence of personhood over the economy *Gaudium et Spes* like Ubuntu states, "persons in extreme necessity are entitled to take what they need from the riches of others."[193] Traditionally, in Sub-Saharan Africa, a starving person, and a stranger has an ethical right to enter into anybody's property and eat. However, he was not allowed to carry spare food from such property. This underlies the precedence of human life over personal or community property. Social justice requires some sort of redistribution to safeguard human dignity of those with disabilities or the marginalized. There is inevitable need to guarantee "the sum total of all those conditions of social life which enable individuals, families, and organizations to achieve complete and effective fulfillment."[194]

The second directive is an explanation for the first one. It reads, "All economic life should be shaped by moral principles. Economic choices and institutions must be judged by how they protect or undermine the life and dignity of the human person, support the family and serve the common good."[195] Once again, this second directive clearly provides ethical scale of priority between personhood and property.

The third directive directly relates economy with morality. It reads, "A fundamental moral measure of any economy is how the poor and vulnerable are faring." This directive results from the increasing marginalization of many poor people due to economic competition which tends to disregard human dignity and morality. In other words, the U.S. bishops are concerned that by ostracizing other human

---

[192] U. S. Catholic Bishops. 1996. "Economic justice for all: A catholic framework for economic life." http://old.usccb.org/jphd/cffel.pdf.

[193] *Gaudium et Spes* (1965, p. 69).

[194] *Gaudium et Spes* (1965, p. 74).

[195] U. S. Catholic Bishops, "Economic justice for all."

persons from the economy the economy becomes immoral and fails to serve its end, which is human being. This directive is inspired by Vatican II document Gaudium et Spes which addresses the "excessive economic and social differences" among local, national, and global populations as a "scandal" which tends to undermine "social justice, equity, the dignity of the human person, as well as social and international peace."[196]

The fourth directive brings to attention the basic human rights. "All people have a right to life and to secure the basic necessities of life (e.g., food, clothing, shelter, education, health care, safe environment, economic security.)"[197] Obviously, those rights imply obligation and necessity for redistribution, people being differently endowed.

The fifth directive of the United States Bishops addresses the right and freedom of active participation that all human beings have. It reads, "All people have the right to economic initiative, to productive work, to just wages and benefits, to decent working conditions as well as to organize and join unions or other associations."

The sixth directive logically flows from the fifth. It states, "All people, to the extent they are able, have a corresponding duty to work, a responsibility to provide the needs of their families and an obligation to contribute to the broader society." This directive cautions against the danger of laxity on the side of the poor. It calls for justice from the poor. The poor do not only need to be protected and be given opportunities to provide for themselves and their families, they need to actively participate in the economic life. In other words, the poor should not take advantage of their situation and fail to make use of their right and responsibility to function as human persons in accordance with the principle of subsidiarity.

The seventh directive calls for governmental active engagement in the economic life. It states, that "In economic life, free markets have both clear advantages and limits; government has essential responsibilities and limitations; voluntary groups have irreplaceable roles, but cannot substitute for the proper working of the market and the just policies of the state."

Closely related to the seventh directive is the eighth directive, which reads, "Society has a moral obligation, including governmental action where necessary, to assure opportunity, meet basic human needs, and pursue justice in economic life."[198] This directive relates to and addresses the situation that John Paul II grieves over: The "innumerable multitude of people—children, adults and the elderly, in other words, real and unique human persons—who are suffering under the intolerable burden of poverty." John Paul II's intention was to draw attention to the alarming marginalization of multitudes of poor people clearly seen in the homelessness, unemployment and international debt.[199]

---

[196] *Gaudium et Spes* (1965, p. 29).

[197] U. S. Catholic Bishops, "Economic justice for all."

[198] U. S. Catholic Bishops, "Economic justice for all."

[199] Pope John Paul II. 1987 Encyclical letter on the concern of the church for the social order *Sollicitudo Rei Socialis*, p. 13.

The ninth and tenth directives are a reminder that human beings involved in economy should remain moral agents, otherwise there is a conflict between economy and personhood. The ninth directive states, "Workers, owners, managers, stockholders and consumers are moral agents in economic life. By our choices, initiative, creativity and investment, we enhance or diminish economic opportunity, community life and social justice."

The tenth directive points to the global economic moral dimension that tends to be forgotten. The directive states, "The global economy has moral dimensions and human consequences." Decisions on investment, trade, aid and development should protect human life and promote human rights, especially for those most in need wherever they might live on this globe.[200] This directive relates with John XXIII's *Pacem in Terris*, which addresses discrimination on the grounds of race, ethnicity, refugee status, class, and gender. John XXIII brings to attention wide discrepancies in economic and social development among global community of states and among individuals. Common good is not limited to one nation; it is a global issue.[201] The directives of the United States Bishops on economic justice for all are based on the primacy of personhood over products of human activity/labor. They are not only in line with St. Thomas' philosophy of the precedence of the first act—that of being, over the second act—that which proceeds from being, they share the worldview of Ubuntu. Human life ought to be always precedent to products of human labor.

5.3.2 The Vulnerable

The second concept of minority empowerment is based on the vulnerable and their right to healthcare. The Catholic Church's tradition of fundamental option for the poor recognizes and affirms the universal right to healthcare.[202] The Church is against abandoning the poor to the vagaries of economic market forces in health care in which only the fittest survive.[203] In and of itself, human dignity justifies healthcare as a basic human right. Health care, therefore, should not be commoditized. According to United States Conference of Catholic Bishops, reliance on market forces has failed to provide a solution to ensuring that the right to care is guaranteed to all by virtue of their common shared humanity.[204]

The bishops argue that those with greater resources should contribute to the health care of others in proportion to their ability to contribute to the financing of

---

[200] U. S. Catholic Bishops, "Economic justice for all."

[201] Pope John XXIII. Encyclical letter on peace on earth *Pacem in Terris*, p. 137.

[202] Pope Leo XIII. Encyclical letter on the new things *Rerum Novarum*; Pius XI. Encyclical letter on in the fortieth year *Quadragesimo Anno*; Pope John Paul II. 1981. Encyclical letter on human labor *Laborem Exercens*. http://www.vatican.va/holy_father/john_paul_ii/encyclicals/documents/hf_jp-ii_enc_14091981_laborem-exercens_en.html.

[203] Baum, Gregory. 1989. "Liberal capitalism," *The logic of solidarity,* 75–89. New York: Orbis.

[204] U. S. Bishops. 1993. "Resolution on health care reform." *Origins* 23(1993): 97, 99–102.

a national health system.[205] Their recommendation interprets and applies Pius XI's encyclical, *Quadragesimo Anno*, that the role of the government includes protecting private individuals and their rights and giving precedence to the poor.[206] The bishops' argument is equally supported by John Paul II's *Sollicitudo Rei Socialis* in which he states that individual claims on the protection of existing resources are not absolute in the face of the more basic needs of others. The pope states categorically that a society dedicated to the pursuit of private property "fails to honor the truth that private property is under a social mortgage."[207]

This papal statement, however, does not deny legitimate control and use of property. It emphasizes that ownership and use of resources must be conditioned by an expansive concern for human wellbeing that views material goods within their ordination to the common good.[208] Human rights of the weak and poor ought to be protected by the government.[209] Workers ought to be protected from being used as tools rather than be treated as ends in themselves.[210]

The society, which is based on the biological principles of struggle for survival and survival of the fittest, is neither humane nor human. The Church's mission in its social teaching is to challenge the society to transcend the principle of survival of the fittest. The Church's teaching on the right to healthcare for the less privileged is contained in the culture of Ubuntu as explicated in Chap. 2. The Church's organized teaching helps enlighten Ubuntu's fundamental worldview on the rights of the poor and under privileged. According to Ubuntu culture, no individual survival is possible independent of the support of the community. Consequently, everybody is vulnerable. An individual is always a part of the whole and needs the whole to be fully realized. Enabling and empowering the vulnerable is an obligation of the fortunate, a right to the vulnerable and a duty of the society to ascertain.

5.3.2.1   Leo XIII on Rights and Duties of Capital and Labor (*Rerum Novarum*)

Although Catholic Church has always delivered social teaching from the time of Christ, the encyclical *Rerum Novarum* of Leo XIII is considered as the summary of the church's position on issues pertaining to justice and social ethics with regards to capital and human labor. Leo XIII was compelled to write his encyclical *Rerum Novarum* by the inhumane condition the working poor were pushed into by their

[205] U. S. Bishops. "Resolution on health care reform," p. 99–100.

[206] Pope Pius XI. Encyclical letter on in the fortieth year *Quadragesimo Anno*, p. 25.

[207] Pope John Paul II. Encyclical letter on the concern of the church for the social order *Sollicitudo Rei Socialis*, p. 42.

[208] Rodger Charles. 1982. *The social teaching of Vatican II*, 299–312. San Francisco: Ignatius Press.

[209] Pope Leo XIII. Encyclical letter on the new things *Rerum Novarum*. http://www.vatican.va/holy_father/leo_xiii/encyclicals/documents/hf_l-xiii_enc_15051891_rerum-novarum_en.html, 54.

[210] Pope Leo XIII. Encyclical letter on the new things *Rerum Novarum*, p. 59.

rich employers. The working poor were in need of rescue from their misery. They were, as well, not protected from any injustice or violence by their employers.[211] The most important requirement on the side of the rich few is to pay justly for the work done by the poor. Since the poor work to keep their lives, live decently and to procure property and better conditions of living, their pay should enable them to do that. Just pay should not be only that which is enough for subsistence. As human beings, not any less than their employers, workers deserve to prosper and enjoy progress and better living as a reward of the work they do.[212]

One of the most important arguments on which Leo XIII bases his argument is the irrefutable fact that in essence as human beings, the poor are not any less than the rich. The poor are equal to the rich also in their human rights and dignity. The poor are as well equal in citizenship with the rich. Moreover, their work is the source of the nation's wealth. It is immoral and ethically unjustifiable, therefore, to belittle or look down on the poor, or judge their human worth based on their lack of property. Religion should not be used to justify the evil of oppression, exploitation and marginalization of the poor.

Moreover, according to Leo XIII, the favor of God seems to incline more towards the poor. Thus, those who consider the poor and minister to them in attitude and deed are God-like.[213] People have an obligation to, as much as they can, liberate the poor from the savagery of greedy people. There are three types of institutions that can help in this noble task. One of the three is working with associations that provide material aid; the second is establishing, enabling and working with privately funded agencies that help workers. The third type of institution is foundations that care for defendants.[214]

Material wellbeing of the poor is not only beneficial for the rich who employ them and the common good of the nation; it is a matter of justice and morality. Being human just like their employers, poor workers deserve enough to enable them to afford a decent shelter, clothes, security and food. Unjust treatment of workers should not be accepted as though it were inevitable. While he encouraged them to stand up for their rights, the Pope discouraged demonstrations and encouraged order.

One should protect one's own rights and interests while refraining from riots and violence. Formation of unions that stand for, and with the poor, is a good idea.[215] A worker should neither intentionally destroy the property of his employer nor forcibly take other persons' properties since the right to personal property must be kept inviolate.[216] Leo XIII urged employers never to treat their employees as means to an end. Being human, they cannot be reduced to a means for other human beings' ends. They are an end themselves. They should neither be treated as slaves nor be used as things for some gain. Since the employer needs his poor worker, it is only wise

---

[211] Pope Leo XIII. Encyclical letter on the new things *Rerum Novarum*, p. 6, 32, 37, 66.

[212] Pope Leo XIII. Encyclical letter on the new things *Rerum Novarum*, p. 9, 55.

[213] Pope Leo XIII. Encyclical letter on the new things *Rerum Novarum*, p. 37, 49, 51.

[214] Pope Leo XIII. Encyclical letter on the new things *Rerum Novarum*, p. 59, 68.

[215] Pope Leo XIII. Encyclical letter on the new things *Rerum Novarum*, p. 37, 51, 54, 69, 82.

[216] Pope Leo XIII. Encyclical letter on the new things *Rerum Novarum*, p. 23, 30, 55.

to have a harmonious humane relationship with him. Mistreating workers is coun-terproductive as it haphazardly affects ownership and business.[217] In sum, workers' human dignity should always be honored. They are persons with physical, spiritual, psychological, moral and familial needs.[218]

The encyclical *Rerum Novarum* is balanced and fair as it seeks justice for all. It urges the poor, for instance, to do the work they are paid for diligently, know-ing that it is not only just, they are actually contributing to the common good as a members of human community.[219] The government should oversee that workers are justly treated and fairly paid. The criterion for discerning fair pay should not be the employers' desires. There must be objective ways to evaluate human labor and reward it.

Wages should minimally be enough to provide for the workers' and their fami-lies' basic needs. If a worker accepts wages which are less than this, he submits to force and violence. Needless to say, work should be reasonable, proportionate and considerate of the workers wellbeing. It should not wear him out or put him into unnecessary risk.[220] The government should oversee employers' treatment of their workers so that workers are not entirely at the mercy of their employ-ers.[221] *Rerum Novarum* is not only substantially in agreement with Ubuntu, it systematically and analytically explains objectives of Ubuntu without intending it. The validity of the philosophy of Ubuntu is vivified by Leo XIII's *Rerum Novarum*. The employer needs the worker as a human being just as the worker needs the employer for the good of both the employer and employee, and for the common good.

5.3.2.2   John Paul II on the Hundredth Year of Rerum Novarum (*Centesimus Annus*)

John Paul II wrote *Centesimus Annus* on the hundredth anniversary of *Rerum No-varum* to assess development based on the challenges and objectives of *Rerum No-varum*. John Paul II not only explores and exposes positive development with re-gards to respect for human dignity, equality and rights, he discovers new challenges confronting the global human community and needing to be reckoned with. One of such problems is Socialism and its tendency to degrade human personhood into a mere element, insignificant relative to national populations.

This error is a serious moral concern since denying any human being his due dig-nity and rights starts a slippery slope in which human rights cannot be defended.[222]

[217] Pope Leo XIII. Encyclical letter on the new things *Rerum Novarum*, p. 28, 31, 32.

[218] Pope Leo XIII. Encyclical letter on the new things *Rerum Novarum*, p. 31, 32.

[219] Pope Leo XIII. Encyclical letter on the new things *Rerum Novarum*, p. 30.

[220] Pope Leo XIII. Encyclical letter on the new things *Rerum Novarum*, p. 59, 61–63, 65.

[221] Pope Leo XIII. Encyclical letter on the new things *Rerum Novarum*, p. 56.

[222] Pope John Paul II. Encyclical letter on the hundredth *Centesimus Annus*. http://www.vatican.va/holy_father/john_paul_ii/encyclicals/documents/hf_jp-ii_enc_01051991_centesimus-annus_en.html, p. 13.

Due to its systematic suppression and undermining of human dignity, rights and freedoms, Socialism is bound to fail and has already been defeated. However, its defeat should not leave Capitalism as the only model of economic organization.[223] In fact, the fall of Marxism highlights human interdependence and mutual neediness.[224]

Human need for other humans is also suppressed, overlooked, or ignored by capitalistic economic systems. There is still marginalization, exploitation and alienation of people, especially in the Third World Countries.[225] On one hand John Paul II cautions the world against exaggerated and uncontrolled capitalistic tendency to let capital dictate the fate of human dignity; on the other hand John Paul II rejoices at the fact that there is growth in the awareness of human rights with United Nations as focal point since 1945.[226] However, the United Nations has not been successful in establishing a development policy or effective system of international conflict resolution that would be a better alternative to war.[227] Although decolonization has already happened, the political liberation has not been true holistic human liberation. There is neo-colonialism, economic manipulation and control of the poor nations by the rich ones. Lack of a competent professional class in the newly independent countries is one of the major factors that perpetuate other forms of colonization via unnecessary dependence.[228] Foreign debt should not be used as a means of undermining, marginalizing and exploiting poor nations; it should be handled in a way that respects human rights to subsistence and progress.[229]

John Paul II notes that there has been a situation of non-war in Europe since the end of the Second World War. One of the observable signs of progress is the collapse of oppressive Eastern Europe regimes in 1989 and the transitioning of some Third World Countries to a more just and participatory structure within themselves and in the global community.[230] However, he observes that non-war is not equivalent to peace. He laments that many people lost, and still don't possess their ability to control their own destiny as arms race and violent extremist groups some with atomic capabilities silently oppress the world.[231] Majority of world populations are structurally denied the basic means to acquire the knowledge they need to enter and participate in the world of technology and intercommunication.[232] They are systematically marginalized. Hence, capitalism still stands in need of addressing its flaws.

---

[223] Pope John Paul II. Encyclical letter on The Hundredth Year *Centesimus Annus*, 35.

[224] Pope John Paul II. Encyclical letter on The Hundredth Year *Centesimus Annus*, 27.

[225] Pope John Paul II. Encyclical letter on The Hundredth Year *Centesimus Annus*, 42.

[226] Pope John Paul II. Encyclical letter on The Hundredth Year *Centesimus Annus*, 21.

[227] Pope John Paul II. Encyclical letter on The Hundredth Year *Centesimus Annus*, 21.

[228] Pope John Paul II. Encyclical letter on The Hundredth Year *Centesimus Annus*, 20.

[229] Pope John Paul II. Encyclical letter on The Hundredth Year *Centesimus Annus*, 35.

[230] Pope John Paul II. Encyclical letter on The Hundredth Year *Centesimus Annus*, 22.

[231] Pope John Paul II. Encyclical letter on The Hundredth Year *Centesimus Annus*, 18.

[232] Pope John Paul II. Encyclical letter on The Hundredth Year *Centesimus Annus*, 33.

Free market economy has failed to satisfy many human needs. There is need for ethics and respect for human rights within economic systems.[233]

John Paul II warns that atheism and contempt for the human person causes class struggle and militarism since peace and human prosperity are natural goods that belong to all members of the human race.[234] Promoting human development of the poor is an opportunity for moral, cultural and economic growth of the entire human race.[235] Observing the principle of subsidiarity is helpful in determining the juridical framework of economic affairs, especially because subsidiarity renders each people an opportunity to participate and achieve self-realization.[236] Development must be holistic by addressing all aspects of humanity, not merely the economic one.[237]

The present generation should always be conscious of their responsibility for the planet for their sake and that of future generations,[238] while, at the same time, creating a lifestyle in which there is quest for truth, beauty, goodness and common good that illumine and help determine choices made.[239] Totalitarianism is an enemy of human progress and it is found in the denial of the transcendental dignity of the human person.[240] Humans need to promote a culture of peace that provide all with realistic opportunities to progress, self-realization and happiness.[241]

Like Ubuntu, John Paul II's *Centesimus Annus* underlines the importance of, not only respecting human dignity, but working together as human family to promote solidarity for the sake of self-realization of all, peaceful community, happiness and meaningful life. Both worldviews guard against exploitation of any human by any other human being due to its effect of depleting the very essence of humanity of meaning. Both views caution against any form of violence against human beings, nature and future generations as contradictory to the meaningful human life and human nature as free and rational beings.

Centesimus Annus is classically, scientifically and systematically written but its perspective is shared by Ubuntu which has been passed on as a philosophy of life based on experience and praxis for many centuries. This philosophy has systematically developed while being enriched by the new challenges that the indigenous Sub-Sahara African communities experienced at any particular time in their history. Thus Ubuntu can be said to validate the practicality and relevance of both *Rerum Novarum* and *Centecimus Annus.* The following section will explore application of the perspective of *Centecimus Annus* on preferential option for the poor from South American liberation theology. Assuming human essential equality which should transcend their accidental endowments.

---

[233] Pope John Paul II. Encyclical letter on The Hundredth Year *Centesimus Annus*, 3–4.

[234] Pope John Paul II. Encyclical letter on The Hundredth Year *Centesimus Annus*, 14, 27.

[235] Pope John Paul II. Encyclical letter on The Hundredth Year *Centesimus Annus*, 28.

[236] Pope John Paul II. Encyclical letter on The Hundredth Year *Centesimus Annus*, 15.

[237] Pope John Paul II. Encyclical letter on The Hundredth Year *Centesimus Annus*, 29.

[238] Pope John Paul II. Encyclical letter on The Hundredth Year *Centesimus Annus*, 37.

[239] Pope John Paul II. Encyclical letter on The Hundredth Year *Centesimus Annus*, 36.

[240] Pope John Paul II. Encyclical letter on The Hundredth Year *Centesimus Annus*, 44.

[241] Pope John Paul II. Encyclical letter on The Hundredth Year *Centesimus Annus*, 52.

5.3.2.3 Gutiérrez on Poverty and Catholic Preferential Option for the Poor

Gustavo Gutiérrez is one of the most known liberation theologians especially following his work, *A Theology of Liberation: History, Politics and Salvation.*[242] In order to capture an overview of his perspective of liberation theology this section is based on an interview found in America: The National Catholic Weekly Titled "Remembering the Poor: An Interview with Gustavo Gutiérrez" of February 3, 2003. According to Gutiérrez, poverty is sub-human condition in which majority of humanity lives today. By describing it as sub-human Gutiérrez implies that it is unfair and unethical and that the minority who are not poor bear some responsibility for the sub-human condition of the majority of their poor counterparts.

Consequently Gutiérrez describes poverty as "more than social issue." It is an ethical issue, a religious issue and a theological issue. Thus "poverty poses a major challenge to every Christian conscience and therefore to theology as well."[243] It cannot be brushed away because it stares at the entire human society, especially those who claim to be Disciples of Christ by trying to follow Christ's footsteps. Gutiérrez confronts the situation of the poor of the world with the Christian message of loving neighbor as oneself. Caring for the neighbor as for oneself is the ideal of Ubuntu. Liberation theology, therefore, is similar to Ubuntu, at least in its ideals. Ubuntu, however, is not Christianity as it is not based on Christian revealed truth. Ubuntu holds that each human being is a unique product of many interconnections facilitated both by God and divinities using other humans and the cosmos.

Thus caring for the poor in Ubuntu is a matter of justice, not charity. Care for the poor, the sick and any needy member of the human family defines the meaning of being human. Thus it is essential to humanity. Gutiérrez's argument for the care of the poor is based on the core teaching of Christ who identified with the poor. Thus, caring for the poor is within the essential substance of Christianity. On the other side, Ubuntu holds that each human being is a unique product of many interconnections facilitated both by God and divinities using other humans and the cosmos. Ubuntu morality can neither ignore interpersonal human relationships nor the necessary existential relationships between human beings and their environment.

Just as humans find themselves in a context, theology is always contextual. Gutiérrez corrects the use of the phrase "contextual theologies" exclusively for liberation theology by stating that theology has always been contextual. "Some theologies, it is true, may be more conscious of and explicit about their contextuality, but all theological investigation is necessarily carried out within a specific historical context."[244] His approach is obviously existential and praxis oriented. Although traditional understanding of theology is that of faith seeking understanding, Gutiérrez focuses on the application of both faith and understanding to human situation rooted in its specific socio-historical and geographical context. Gutiérrez describes our present context as one that is "characterized by a glaring disparity between the rich

---

[242] Gutiérrez (1973).

[243] Hartnett (2003).

[244] Hartnett (2003).

and the poor." It is this disparity which is at the base of liberation theology movement, since it has to be addressed.[245] In other words, Gutiérrez posits that a theology that abstracts from reality while ignoring the factual, spacio-temporal reality is unrealistic.

Gutiérrez observes that "Poverty has a visibility today that it did not have in the past. The faces of the poor must now be confronted." Christians and the entire human society cannot and ought not to ignore the reality of poverty that confronts them, demanding action and solution. There is simply no escape. "No serious Christian can quietly ignore this situation." Ignorance of the plight of the poor is not invincible. Fortunately, nowadays there are means to resolve the problem of poverty. In the distant past "poverty was considered to be an unavoidable fate, but such a view is no longer possible or responsible" because now human society has the ability to ascertain provision of the decent minimum of human needs for dignified life.

More importantly "we also understand the causes of poverty and the conditions that perpetuate it. Now we know that poverty is not simply a misfortune; it is an injustice."[246] Even though Gutiérrez does not qualify or define his use of the word 'injustice' in this case, there certainly is some injustice in the denial of dignity and means to sustain that dignity to every poor human being. Nobody is exempt from this kind of omission.

Gutiérrez describes material poverty as "premature and unjust death" since the poor are forced by their situation to succumb to the consequences of their poverty. They are thus denied of their human dignity, either actively or passively by their fellow humans. In reality a poor person is thus being reduced or treated "as a non-person, someone who is considered insignificant from an economic, political and cultural point of view." In other words, in the judgment of other human beings "the poor count as statistics; they are the nameless." However, in the sight of God, the poor are "never insignificant." As humans they are equal to the rich.[247]

According to Gutiérrez, therefore, "the option for the poor is not optional, but is incumbent upon every Christian. It is not something that a Christian can either take or leave. As understood by Medellín, the option for the poor is twofold: it involves standing in solidarity *with* the poor, but it also entails a stance *against* inhumane poverty."[248] Hence option for the poor is a task, a duty and responsibility that is shared by all human beings, especially Christians.

Gutiérrez boldly states that "option for the poor has become part of the Catholic social teaching. The phrase comes from the experience of the Latin American church. The precise term was born sometime between the Latin American bishops' conferences in Medellín (1968) and in Puebla (1979)." However, as a theology, it is universal. "The content, the underlying intuition, is entirely biblical. Liberation theology tries to deepen our understanding of this core biblical conviction."[249] This

---

[245] Hartnett (2003).
[246] Hartnett (2003).
[247] Hartnett (2003).
[248] Hartnett (2003).
[249] Hartnett (2003).

conviction is always rooted in the actual situation on a specific ground "a good contextual theology, though, will also deal with global issues, because Christian responsibility does not stop at the border. The ministry of solidarity has international dimensions."[250] In other words, being human is not exclusive of other humans regardless their location. There is a basic connection between all human beings that hold all responsible for all others.

Preferential option for the poor is part and parcel of Catholic socio-ethical teaching. It started with Christ himself. It runs through most encyclicals and Vatican II Documents. In his Apostolic Exhortation *Evangelii Nuntiandi*, Pope Paul the VI states that it is the duty of the church "to proclaim the liberation of millions of human beings ... struggle to overcome everything which condemns them to remain on the margin of life: famine, chronic disease, illiteracy, poverty, injustices in international relations and especially in commercial exchanges, situations of economic and cultural neo-colonialism sometimes as cruel as the old political colonialism."[251]

John Paul II has been on the forefront in defense of preferential option for the poor. In his encyclical *Sollicitudo rei Socialis* he states "the option or love of preference for the poor ... a special form of primacy in the exercise of Christian charity, to which the whole tradition of the Church bears witness."[252] In his *Centesimus Annus*, John Paul II categorically states, "love for others, and in the first place love for the poor, in whom the Church sees Christ himself, is made concrete in the promotion of justice."[253] In *Sollicitudo rei Socialis*, Pope John Paul II says that solidarity is "not a feeling of vague compassion or shallow distress at the misfortunes" of the poor but "a firm and persevering determination to commit oneself to the common good."[254]

In sum, Catholic socio-ethical tradition has always been in defense of the poor; and for their liberation and protection. Like the Catholic socio-ethical tradition, UNESCO Declaration on Bioethics and Human Rights, though it arrived a little late to the scene, it is in defense for the poor. In many ways Ubuntu identifies both Catholic socio-ethical teaching, liberation theology and UNESCO Declaration on Bioethics and Human Rights. One of the major ways Ubuntu does it practically is its refusal to let any human being fall below what is acceptable by the society as decent enough for his dignity as human.

There is always an imaginary poverty line in the cultures that share Ubuntu worldview. Such line is relative to average wealth within the society. Ubuntu's worldview is praxis oriented. For the sake of the human essence every member of the society participates, every person is actively responsible and engaged in the

---

[250] Hartnett (2003).

[251] Pope Paul VI. 1975. Apostolic exhortation on evangelization in the modern world *Evangelii Nuntiandi*. http://www.vatican.va/holy_father/paul_vi/apost_exhortations/documents/hf_p-vi_exh_19751208_evangelii-nuntiandi_en.html, p. 30.

[252] Pope John Paul II. 1987. Encyclical letter on the concern of the church for the social order *Sollicitudo Rei Socialis*, p. 42.

[253] Pope John Paul II. 1991. Encyclical letter on the hundredth year *Centesimus Annus*, p. 58.

[254] Pope John Paul II. 1987. Encyclical letter on the concern of the church for the social order *Sollicitudo Rei Socialis*, p. 38.

plight of the poor. Ubuntu is proactive with regards to this issue since its stance is that of prevention. According to Ubuntu, no one is truly free from the plight of any marginalized human being.

5.4   Conclusion

The Church's teaching emphasizes that human dignity is inherent in each human being. It is the inherent human dignity in itself that commands recognition and respect. Respecting human dignity means observing and protecting human rights. Like Catholic tradition, the culture of Ubuntu has deep respect for human dignity. In both Catholic socio-ethical teaching and Ubuntu worldview, a human person is a unique beginning and end. Each person commands attention, respect and dignity worth his nature as unrepeatable unique event. Theological truths about human personhood should never be left aside in any realistic human development. Any attempt to do that leads to absurdity. Due to human dignity and rights both Catholic teaching and Ubuntu respect Common good without violating individual freedoms and rights. Common good in practice means that the society has to assure human dignity by ascertaining the decent minimum of care for all. The society provides security and assurance of protection to its members.

The teaching of the importance of human community in the Catholic Catechism is replicated in the Ubuntu worldview. The only substantial exception is Catholic's religious dimension of human community founded on, and inspired by, the Sacred Trinity. Even though Ubuntu ideal of community and society is religious, it is not explicitly Christian. It can be argued, however, that Ubuntu worldview is a kind of anonymous Christianity due to Ubuntu worldview's substantial resemblance to Christian Ideals.

Ubuntu understanding of solidarity finds endorsement in the Catholic traditional social teaching. That is, nobody should be marginalized since doing so destroys not just the victim but also the offender and the entire human species. Ostracizing a human person depletes the very essence of humanity of meaning. Self-realization of humans, whether moral, cognitive or sociological, is never independent of other human beings. This essential contingency of humans to fellow humans is the core of Ubuntu worldview. Human beings need other human beings, those currently alive, those who have passed on, on whose shoulders the present generation stands, and those to come. The present generation needs future generations to continue what the present generation has received and started, or else it is all absurd and meaningless.

Nobody can live for oneself alone. Mutuality is characteristic of human beings. Human actions have dual characteristics, that is, at once for self and for the society. The greatest teacher and inspiration of this fact is Christ who gave of himself for the salvation of the human race. On the side of Ubuntu the inspiration comes from the awareness that all that a person is, is received and what he is and have, have to be shared or passed on. Catholic socio-ethical tradition has always been in defense of the poor; and for their liberation and protection. In many ways, though unconsciously, Ubuntu identifies both with Catholic socio-ethical teaching and liberation theology.

One of the major ways Ubuntu does it practically is in its refusal to let any human being fall below poverty line, relative to the community's economy. The poor in the society become the concern of every member of that specific society. The gap between the poor and the rich is the yardstick that measures morality of both the society and the individuals' in it. Catholic socio-ethical teaching is normally first documented before it is put to practice; Ubuntu teaching is praxis oriented, based on centuries of experience that has been handed over. Ubuntu is a communitarian ethic. Every member of the society participates in the struggle against all that is contrary to both individual and communal good; consequently, every person is actively responsible and engaged in the plight of the poor both at an individual level and at a societal level. Ubuntu is proactive since its stance is that of prevention. Ubuntu cautions, not only against dehumanizing poverty but also against disproportional riches. Ubuntu recognizes the fact that the reason there are excessively rich people in the society is that there are excessively poor people. Ubuntu watches against that as a matter of faith and morality.

Conclusion

This book interprets the culture of Ubuntu to explain the contribution of a representative indigenous African ethics to global bioethics. Specifically, Ubuntu presents a representative communal worldview for ethical decisions whereby individuals, community and world are connected together. Ubuntu ethics protects individual rights within a cosmic context to enhance solidarity. However, solidarity is essential for maximization of quantity and quality of human life. Precisely, Ubuntu worldview promotes life-centered ethics.

Specifically, this work demonstrates that Ubuntu is a representative world view that upholds respect for persons, construed in terms of their dignity and rights, in the context of relationality with the cosmos and the subsidiarity with the human community. Human relationships are important because they help generate, recognize, promote and nurture human life. Human relationships with the biosphere and the cosmos ought to be life nurturing, life maximizing and life promoting mainly because human life is dependent on its immediate environment and the cosmos.

Ubuntu holds that maintenance of optimal equilibrium, integrity and sanctity of the cosmos is a sacred and moral obligation. Humans have a duty and obligation to provide good stewardship, treasure, and safeguard their environment both for the current and for future generations as a matter of ethics. Future generations belong to the realm of "the other" without whom "the self" cannot be defined. Cognition and its development does not happen independent of acknowledgement of otherness. Equally, moral development is other—centered. Thus Ubuntu worldview contrasts Cartesian's *Cogito Ergo Sum*. Every member of the human genre has an ultimate personal obligation to grow, that is, to become fully human. To become fully human means to maximize both personal and communal life by increasingly entering into community with other persons and the cosmos without losing or compromising one's individuality.

Healthcare in Ubuntu is a concern of all members of society. From the perspective of Ubuntu, the poor and the underprivileged have a just claim to the labor, talent and time of the community in whose life they share. It is a moral duty, not optional charity, to provide for those who cannot provide for themselves while recognizing and appreciating their contribution, according to the principle of subsidiarity. No human life is in vain. When human life is at stake, no individual rights holds. Hu-

L. T. Chuwa, *African Indigenous Ethics in Global Bioethics,* Advancing Global Bioethics, 245
DOI 10.1007/978-94-017-8625-6, © Springer Science+Business Media Dordrecht 2014

man life overrides all individual rights, except when such life is a threat to more lives or the life of the community. Caring is a proof of both personal and community moral maturity. Ubuntu assumes that the welfare of individuals is dependent on the welfare of the community as a whole, just as it assumes that 'being an individual is being with others' and that the self stands in constant need of an-other.

Hence the individual does not take precedence over the community. Initiations are geared toward acknowledgement that ethically, individual rights meet their limit in the rights of other individuals represented in sum by the community. It is the continual process of initiation into and through the community, which enables sub-Sahara Africans to think in 'both/and rather than either/or' categories. Personal maturity is realized in the process of synthesizing and reconciling of personal autonomy with other persons' autonomy. One of the basic functions of the society is to ascertain and protect human relationships so that they promote and nourish life. Consequently, Ubuntu synthesizes the tension between the ethical principles of autonomy, justice and beneficence.

Ubuntu understands human disease comprehensively, essentially as a breach or breakage of human integrity. Ubuntu healthcare addresses not only the visible symptoms, but the possible underlying physiogenic, psychological, social and ontological causes. Healing, therefore, is a process of reconciliation. Healing reconciles and restores the lost unity within the self; between the self and the society; between the self and the diseased; between the self and the cosmos; and between the self and God. Thus, Ubuntu perspective on human disease and healing is comprehensive and holistic. Essentially, Ubuntu is undeniably an ethic of care that conforms to the ideals and objectives of the UNESCO Declaration on Bioethics and Human Rights. Even though Ubuntu is not explicitly Christian, its substance makes it an anonymous Christian worldview, especially due to its shared perspective with Christian social ethics.

The first part (Chaps. 1, and 2) explains the meaning of Ubuntu, as a representative indigenous ethics, and its three constituent components. This part demonstrates that Ubuntu represents indigenous ethics that focus on the centrality of respecting rights based on human dignity, recognizing the cosmic context of ethics, to enhance the role of solidarity.

The second part (Chaps. 3, 4, and 5) interprets the culture of Ubuntu as providing a representative African ethics that contributes to global bioethics. To explore this contribution to global bioethics, the three components of Ubuntu ethics are analyzed in light of major approaches to contemporary bioethics discourse.

In Chap. 3, the component of rights in Ubuntu ethics is explored as being consistent with and enhancing bioethics discourse in the ethics of care. In Chap. 4 the component of cosmic context is explored as being consistent with and enhancing bioethics discourse related to the UNESCO Declaration on Bioethics and Human Rights. In Chap. 5 the component of solidarity in Ubuntu ethics is explored as being consistent with and enhancing bioethics discourse in the global Catholic tradition of social ethics.

In sum Ubuntu makes a valid contribution to global bioethics that ought to be seriously considered. The contribution that Ubuntu makes to global bioethics is paradoxically facilitated by its openness to systematized and principled enlightenment by major global trends in bioethics such as Ethics of Care, UNESCO Declaration on Bioethics and Human Rights and the Catholic Traditional Socio-Ethical Teaching. All major trends in ethics presume and share in the assertion of Ubuntu that the very essence of the human "*self*" is annihilated by the sheer absence of the "*other*." Selfhood is undeniably and helplessly dependent on otherness. Ignorance of otherness is ignorance of the selfhood.

Bibliography

Aday, Lu Ann. 1993. *At risk in America: The health and health care needs of vulnerable populations in the United States.* San Francisco: Jossey-Bass.

Agius, Emmanuel, and Savino Busuttil. 1998. *Germ-line intervention and our responsibilities to future generations.* Dordrecht, The Netherlands: Kluwer Academic Publishers.

Allison, Kirk. 2005. UN bioethics and cloning declarations overlap. SciDev.Net. http://www.scidev.net/EditorLetters/index.cfm?fuseaction=readeditorletter&itemid=80language=1. Accessed 8 July 2012.

Allmark, Peter. 1995. Can there be an ethics of care? *Journal of Medical Ethics* 21:19–24.

Amstutz, Mark R. 2008. *International ethics: Concepts, theories, and cases in global politics.* New York: Bowman and Littlefield Publishers.

Andorno, Roberto. 2007. Global bioethics at UNESCO: In defense of the universal declaration on bioethics and human rights. *Journal of Medical Ethics* 33:150–154.

Andorno, Roberto. 2009a. April 22. Human dignity and human rights as a common ground for a global bioethics. *Journal of Medicine and Philosophy* 34:235–240.

Andorno, Roberto. 2009b. Human dignity and human rights. In *The UNESCO universal declaration on bioethics and human rights: Background, principles and application,* eds. Henk A. M. J. ten Have and Michele S. Jean, 91–98. Paris: United Nations Educational and Cultural Organization.

Aristotle. 2000. *Nicomachean ethics,* ed. Roger Crisp Cambridge. United Kingdom: Cambridge University Press.

Asante, Molefi Kete, Yoshitaka Miike, and Jing Yin, eds. 2008. *The global intercultural communication reader.* New York: Routledge.

Battle, Michael. 1997. *The Ubuntu theology of Bishop Desmond Tutu.* Cleveland: Pilgrim's Press.

Battle, Michael. 2009. *Ubuntu: I in you and you in me.* New York: Seabury Books.

Baum, Gregory, and Robert Ellsberg, eds. 1989. *The logic of solidarity: Commentaries on Pope John Paul II's encyclical on social concern.* Maryknoll: Orbis.

Beauchamp, Dan E. 1988. *The health of the republic: Epidemics, medicine, and moralism as challenges to democracy.* Philadelphia: Temple University Press.

Beauchamp, Dan E., and James F. Childress. 2009. *Principles of biomedical ethics.* New York: Oxford University Press.

Benatar, S. R. 1998. Global disparities in health and human rights: A critical commentary. *American Journal of Public Health* 88:295–300.

Benatar, David. 2005. The trouble with universal declarations. *Developing World Bioethics* 5 (3): 220–224.

Benhabib, Seyla. 1997. *Situating the self: Gender, community and postmodernism in contemporary ethics.* New York: Routledge.

Berg, A. 2003, June. Ancestor reverence and mental health in South Africa. *Transcultural Psychiatry* 40 (2): 194–207.

Bettistella, R. M., and J. M. Kuder. 1993. Universal access to health care: A practical perspective. *Journal of Health and Human Resources Administration* 16:6–34.

Bhengu, M. J. 1996. *Ubuntu: The essence of democracy*. Cape Town: Novalis Press.

Biko, Steve. 2004. *I write what I like*. Johannesburg: Picador Africa.

Boyle, P. J., et al. 2000. *Organizational ethics in health care*. San Francisco: Jossey-Bass.

Briney, Amanda. 2011. An overview and history of UNESCO. http://geography.about.com/od/politicalgeography/a/unesco.htm. Accessed 25 Sept 2013.

Broodryk, Johann. 1997. *Ubuntu management and motivation*. Johannesburg: Gauteng Department of Welfare.

Brown, Lee M., ed. 2004. *African philosophy: New and traditional perspectives*. New York: Oxford University Press.

Buber, Martin. 1958. *I and Thou*. 2nd ed. New York: Charles Scribner's Sons.

Buchanan, Allen. 1984. The right to a decent minimum of health care. *Philosophy and Public Affairs* 13 (1): 55–78.

Budge, E. A. Wallis. 1895. *The Egyptian book of the dead: (The Papyrus of Ani) Egyptian text transliteration and translation*. New York: Dover Publications, Inc Trustees of the British Museum.

Bujo, Bénézet. 1992. *African theology in its social context*. Eugene: Wipf and Stock Publishers.

Bujo, Bénézet. 1998. *The ethical dimension of community. The African model and the dialogue between North and South*. Nairobi: Paulines Publications Africa.

Bujo, Bénézet. 2001. *Foundations of African ethic: Beyond the universal claims of western morality*. New York: The Crossroad Publishing Company.

Callahan, Daniel, and Angela A. Wasunna. 2006. *Medicine and the market: Equity v. choice*. Baltimore: The John Hopkins University Press.

Catechism of the Catholic Church. 1994. 1st ed. Washington, D.C: Liguori Publications.

Catechism of the Catholic Church. 1997. 2nd ed. Vatican web site. http://www.vatican.va/archive/ccc_css/archive/catechism/ccc_toc.htm. Accessed 14 March 2012.

Centers for Disease Control and Prevention. National center for health statistics. http://www.cdc.gov/nchswww. Accessed 3 March 2012.

Chachine, Isaias Ezekiel. 2008. *Community, justice, and freedom: Liberalism, communitarianism, and African contributions to political ethics*. Sweden: Edita Vastra Aros, Vasteras.

Chen, Guo-Ming, and William J. Starosta. 2008. Intercultural communication competence: A synthesis. In *The global intercultural communication reader*, eds. Molefi Kete Asante, Yoshitaka Miike, and Jing Yin, 215–230. New York: Routledge.

Cherry, Mark J. 2009, November 13. Religion without god, social justice without christian charity, and other dimensions of the culture wars. *Christian Bioethics* 15 (3): 277–299.

Childress, James F., and John Macquarie, eds. 1986. *The westminster dictionary of christian ethics*. Philadelphia: The Westminster Press.

Chodorow, Nancy. 1978. *The reproduction of mothering*. Berkeley: University of California Press.

Christiansen, Drew. 1991. The great divide: Catholic social teaching and American health care. *Linacre Quarterly* 58:40–49.

Chuwa, Leonard Tumaini. 2013. Interpreting the culture of Ubuntu: The contribution of a representative indigenous African ethics to global bioethics. http://digital.library.duq.edu/cdm-etd/document.php?CISOROOT=/etd&CISOPTR=154279&REC=9.

Clement, Grace. 1996. *Care, autonomy, and justice: Feminism and the ethics of care*. Boulder: Westview Press.

Coetzee, P. H., and A. P. J. Roux, eds. 2003. *The African philosophy reader*. New York: Routledge.

Colby, Anne, J. Gibbs, Marcus Lieberman, and Lawrence Kohlberg. 1983. *Longitudinal study of moral judgment: A monograph for the society of research in child development*. Chicago: The University of Chicago Press.

Coleman, Carl H., Marie-Charlotte Bouësseaub, and Andreas Reisb. 2008, August. The contribution of ethics to public health. *Bulletin of the World Health Organization* 86 (8). http://www.who.int/bulletin/volumes/86/8/08-055954/en/index.html. Accessed 7 Aug 2012.

Conn, Marie A. 1997. Health-care reform: A human rights issue. In *life and learning. VI Proceedings of the sixth University faculty for life conference*, 899–1000. Washington D.C.: University Faculty for Life.

Conneen, Chris. 2008. Exploring the relationship between reparations, the gross violation of human rights and restorative justice. In *Handbook of restorative justice*, eds. Sullivan Dennis and Tifft Larry, 355–368. New York: Routledge.

Cox Macpherson, Cheryl. 2007. October. Global bioethics: Did the universal declaration on bioethics and human rights miss the boat? *Journal of Medical Ethics* 33 (10): 588–590.

Daniels, Norman. 1985. *Just health care*. Cambridge University Press.

Daniels, Norman. 2008. *Just health: Meeting health needs fairly*. New York: Cambridge University Press.

d'Empire, Gabriel. 2009. Equality, justice and equity. In *The UNESCO universal declaration on bioethics and human rights: Background, principles and application,* eds. Henk A. M. J. ten Have and Michele S. Jean, 173–185. Paris: United Nations Educational and Cultural Organization.

De Castro, Leonardo D., Peter A. Sy, and Teoh Chin Leong. 2011. Poverty and indigenous peoples. In *The Sage handbook of health care ethics,* ed. Ruth Chadwick, Henk ten Have, and Eric M. Meslin, 289–305. Thousand Oaks: Sage.

De Chardin, Pierre Teilhard. 1969. *Human energy*. London: William Collins Sons and Co. Ltd.

Derrida, Jacques. 1990. Force of law: The 'mystical foundation of authority.' *Cordozo Law Review* 11 (5–6): 919–1045.

Dewey, John. 1929. *Democracy and education*. New York: MacMillan.

Dickson, David. 2005. Sparks fly over UNESCO bioethics pact. SciDev.Net. http://www.scidev.net/News/index.cfm?fuseaction=readnews&itemid=2433&language=1. Accessed 7 Aug 2012.

Donnelly, Jack. 1994. International human rights and health care reform. In *Health care reform: A human rights approach,* ed. Audrey R. Chapman, 134–135. Washington, D.C: Georgetown University Press.

Dorr, Donal. 1989. Solidarity and integral human development. In *The logic of solidarity,* ed. Baum Gregory and Ellsberg Robert, 48–49. New York: Orbis.

du Pre, Athena. 2000. *Communication about health care: Current issues and perspectives*. Mountain View: Mayfield.

du Toit, B. 1980. Religion, ritual, and healing among urban black South Africans. *Urban Anthropology* 9 (1): 21–49.

Emergency Medical Treatment and Active Labor (EMLATA), 42 U.S. Congress (1998).

Emmet, Dorothy. 1979. *The moral prism*. New York: St. Martin's Press.

Etheredge, Lynn, Stanley B. Jones, and Lawrence Lewin. 1996. What is driving health system change? *Health Affairs* 15 (4): 93–104.

Faunce, Thomas Alured. 2011. Emerging technologies: Challenges for health care and environmental ethics and rights in an era of globalization. In *The sage handbook of health care ethics,* ed. Ruth Chadwick, Henk ten Have, and Eric M. Meslin, 49–62. Thousand Oaks: Sage.

Finnis, John. 1980. *Natural law and natural rights*. New York: Oxford University Press.

Freudenheim, M. 1999. Employees face steep increases in health care. *The New England Journal of Medicine* 340 (3): 248–252.

Fried, Charles. 1976. Equality and rights in medical care. *Hastings Center Report* 6:29–34.

Garrafa, Volnei, et al. 2010. Between the needy and the greedy: The quest for a just and fair ethics of clinical research. *Journal of Medical Ethics* 36:500–504.

Gbadegesin, S. 1991. *African philosophy: Traditional Yoruba philosophy and contemporary African realities*. New York: Peter Lang.

Giles-Vernick, T., and S. Rupp. 2006. Visions of apes, reflections on change: Telling tales of great apes in equatorial Africa. *African Studies Review* 49 (1): 61–62.

Gilligan, Carol. 1982. *In a different voice: Psychological theory and women's development*. Cambridge: Harvard University Press.

Gilligan, Carol. 1987. Moral orientation and moral development. In *Women and moral theory,* ed. Kittay and Meyers, 19–33. Totowa: Rowman & Litttlefield.

Grimshaw, Jean. 1986. *Philosophy and feminist thinking*. Minneapolis: University of Minnesota Press.

Gunson, Darryl. 2009, April 22. Solidarity and the universal declaration on bioethics and human rights. *Journal of Medicine and Philosophy* 34 (3): 241–260.

Gutiérrez, Gustavo. 1973. *A theology of liberation: History, politics and salvation*. Maryknoll: Orbis Books.

Gyekye, Kwame. 1997. *Tradition and modernity: Philosophical reflections on the African experience*. New York: Oxford University Press.

Hartnett, Danielle. 2003, February 3. Remembering the poor: An interview with Gustavo Gutiérrez. *American Magazine*. http://www.americamagazine.org/content/article.cfm?article_id=2755. Accessed 10 March 2012.

Held, Virginia. 2006. *The ethics of care: Personal, political and global*. New York: Oxford University Press.

Hessler, Kristen, and Allen Buchanan. 2002. Specifying the content of the human right to health care. In *Medicine and social justice: Essays on the distribution of health care*, ed. Rosamond Rhodes, Margaret P. Battin, and Anita Silvers, 84–95. New York: Oxford University Press.

History of the United Nations. 2013. http://www.un.org/en/aboutun/history/. Accessed 25 Sept 2013.

Holdstock, T. Len. 2000. *Re-examining psychology: Critical perspectives and African insights*. London: Routledge.

Hollenbach, David. 1979. *Claims in conflict: Retrieving and renewing the catholic human rights tradition*. New York: Paulist Press.

Hollenbach, David. 1989. The common good revisited. *Theological Studies* 50:70–94.

Hord, Fred Lee, and Jonathan Scott Lee. 1995. *I am because we are: Readings in black philosophy*. Amherst: University of Massachusetts Press.

Institute of Medicine. 2002. *Coverage matters: Insurance and health care*. Washington, D.C.: National Academies Press.

Institute of Medicine. 2003, June. *Hidden costs, value lost: Unisurance in America*. Washington, D.C.: National Academies Press.

Jean, Michelle. 2004, December. Human dignity, key value of bioethics. *The New Courier* 4–6. (Interview conducted by Jeanette Blom).

Johnson, R. W., and David Welsh, eds. 1998. *Ironic victory: Liberalism in post-liberation South Africa*. New York: Oxford University Press.

Jonsen, Albert. 1998. *The birth of bioethics*. New York: Oxford University Press.

Kalfoglou, Andrea. 2011. Reprogenetics. In *The Sage handbook of health care ethics*, eds. Ruth Chadwick, Henk ten Have, and Eric M. Meslin, 179–193. Thousand Oaks: Sage.

Kamalu, C. 1997. *Person, divinity and nature: A modern view of the person and the cosmos in African thought*. London: Kamak House.

Kamwangamalu, Nkonko M. 2008. Ubuntu in South Africa: A sociolinguistic perspective to a Pan-African concept. In *The global intercultural communication reader*, ed. Molefi Kete Asante, Yoshitaka Miike, and Jing Yin, 113–122. New York: Routledge.

Kasenene, Peter. 1994. Ethics in African theology. In *Doing ethics in context: South African perspectives*, ed. C. Villa-Vicencio and J. W. de Gruchy, 139–147. Cape Town: David Philip Publishers.

Kelsen, Hans. 1996. What is justice? In *Justice*, ed. Jonathan Westphal, 183–206. Indianapolis: Hackett Publishing.

Kinoti, H. 1999. African morality: Past and present. In *Moral and ethical issues in African christianity*, ed. J. Mugambi and A. Nasimiyu-Wasike, 73–82. Nairobi: Action Publishers.

Kohlberg, Lawrence. 1958. The development of modes of moral thinking and choice in the years ten to sixteen. Ph. D. Dissertation, University of Chicago.

Kohlberg, Lawrence. 1963a. The development of children's orientation toward a moral order: I. Sequence in the development of moral thought. *Vita Humana* 6:11–33.

Kohlberg, Lawrence. 1963c. Psychological analysis and literary form: A study of the doubles in Dostroevsky. *Daedalus* 92:345–363.

Kohlberg, Lawrence. 1968a. The child as a moral philosopher. *Psychology Today* 1:25–30.

Kohlberg, Lawrence. 1968b. Early education: A cognitive-development view. *Child Development* 39:1013–1062.

Kohlberg, Lawrence. 1969. Stage and sequence: The cognitive-developmental approach to socialization. In *Handbook of socialization theory and research,* ed. David A. Goslin, 443–480. Chicago: Rand McNally & Company.

Kohlberg, Lawrence. 1971a. Stages of moral development as a basis for moral education. In *Education: Interdisciplinary approaches,* ed. C. M. Beck, B. S. Crittenden, and E. V. Sullivan. Toronto: University of Toronto Press.

Kohlberg, Lawrence. 1971b. *From is to ought: How to commit the naturalistic fallacy and get away with it in the study of moral development.* New York: Academic Press.

Kohlberg, Lawrence. 1981. *Essays on moral development, vol. I: The philosophy of moral development.* San Francisco: Harper & Row.

Kohlberg, Lawrence, and T. Lickona, ed. 1976. Moral stages and moralization: The cognitive developmental approach. In *Moral development and behavior: Theory, research and social issues,* 31–53. Holt: Rinehart and Winston.

Kohlberg, Lawrence, Charles Levine, and Alexandra Hewer. 1983. *Moral stages: A current formulation and a response to critics.* New York: Karger.

Korthals, Michael. 2011. Ethics of environmental health. In *The Sage handbook of health care ethics,* ed. Ruth Chadwick, Henk ten Have, and Eric M. Meslin, 413–426. Thousand Oaks: Sage.

Kuttner, Robert. 1999a. Employer-sponsored health coverage. *The New England Journal of Medicine* 340 (3): 248–252.

Kuttner, Robert. 1999b. Health insurance coverage. *The New England Journal of Medicine* 340 (2): 163–168.

Langlois, Adele. 2008, June. The UNESCO declaration on bioethics and human rights: Perspectives from Kenya and South Africa. *Health Care Annals* 15:39–51.

Legum, Colin, and Geoffrey Mmari. 1995. *Mwalimu: The right influence of Nyerere.* Trenton: African World Press.

Leo XIII. 1891, May 15. Rerum Novarum. Encyclical letter on capital and labor. Vatican web site. http://www.vatican.va/holy_father/leo_xiii/encyclicals/documents/hf_l-xiii_enc_15051891_rerum-novarum_en.html. Accessed 12 June 2012.

Louw, Dirk J. 2007. Ubuntu: An African assessment of the religious other. Paper presented at the Twentieth World Congress of Philosophy. http://www.bu.edu/wcp/papers/Afri/AfriLouw.htm. Accessed 11 May 2012.

MacIntyre, Alasdair. 1984. *After virtue.* 2nd ed. Notre Dame: University of Notre Dame, Indiana.

Mackler, Aaron L. 1991, June. Judaism, justice, and access to health care. *Kennedy Institute of Ethics Journal* 1 (2): 143–161.

Macquarrie, John. 1972. *Existentialism.* London: Penguin Books.

Magesa, L. 1997. *African religion: The moral traditions of abundant life.* New York: Orbis Books.

Makinde, M. Akin. 1988. *African philosophy, culture, and traditional medicine.* Athens: Center for International Studies Ohio University.

Mandela, Nelson. 1994. *A long walk to freedom: The autobiography of Nelson Mandela.* Boston: Little, Brown and Company.

Marquard, Leo, and T. G Standing. 1939. *The Southern Bantu.* London: Oxford University Press.

Martin, Kelley. Top 5 causes of World War I. http://americanhistory.about.com/od/worldwari/tp/causes-of-world-war-1.htm. Accessed 25 Sept 2013.

Masolo, D. A. 1994. *African philosophy in search of identity.* Bloomington: Indiana University Press.

Mbiti, John S. 1990. *African religions and philosophies.* New Hampshire: Heinemann Educational Books Inc.

Mbombo, O. 1996. Practicing medicine across cultures: Conceptions of health, communication and consulting practice. In *Cultural synergy in South Africa. Weaving strands of Africa and Europe,* ed. M. Steyn and K. Motshabi. Randburg: Knowledge Resources.

McDonald, Michael, and Nina Preto. 2011. Health research in the global context. In *The Sage handbook of health care ethics,* ed. Ruth Chadwick, Henk ten Have, and Eric M. Meslin, 327–329. Thousand Oaks: Sage.

Menkiti, Ifenyi. 1984. Person and community in African thought. In *African philosophy: An introduction,* ed. R. Wright, 170–175. Lanham: University Press of America.

Metz, Thaddeus. 2004, November. An African theory of moral status: A relational alternative to individualism and holism. *Ethical theory and moral practice: An alternative forum* 7 (5). http://www.filosofi.uu.se/digitalAssets/110/110279_thaddeus-metz-an-african-theory-of-moral-status.pdf. Accessed 8 Aug 2012.

Metz, Thaddeus. 2007. Toward an African moral theory. *The Journal of Political Philosophy* 15 (3): 321–344.

Metz, Thaddeus. 2009. The final ends of higher education in light of an African moral theory. *Journal of Philosophy of Education* 43 (2): 179–201.

Metz, Thaddeus. 2010a. An African theory of bioethics: Reply to MacPherson and Maclin. *Developing World Bioethics* 10 (3): 158–163.

Metz, Thaddeus. 2010b. Human dignity, capital punishment, and an African moral theory: Toward a new philosophy of human rights. *Journal of Human Rights* 9 (1): 81–99. http://www.tandfonline.com/loi/cjhr20. Accessed 23 April 2012.

Metz, Thaddeus, and Joseph B. R. Gaie. 2010. The African ethic of Ubuntu/Botho: Implications for research on morality. *Journal of Moral Education* 39 (3): 273–290. http://www.tandfonline.com/loi/cjme20. Accessed 8 Aug 2012.

Mnyaka, Mleki, and Mokgethi Motlhabi. 2003. The African concept of Ubuntu/Batho and its socio-moral significance. *Black Theology: An International Journal* 3 (2): 224.

Morisaki, Takayuki. 2009. Protecting future generations. In *The UNESCO universal declaration on bioethics and human rights: Background, principles and application,* ed. Henk A. M. J. ten Have and Michele S. Jean, 243–245. Paris: United Nations Educational and Cultural Organization.

Mosala, Itumeleng J., and Buti Tilhagale, ed. 1986. *The questionable right to be free: Black theology from South Africa.* Maryknoll: Orbis Books.

Mueller, Franz H. 1943. The principle of subsidiarity in the christian tradition. *The American Catholic Sociological Review* 4:147–157.

Murove, Felix Munyaradzi. 2004, winter. An African commitment to ecological conservation: The Shona concepts of Ukama and Ubuntu. *The Mankind Quarterly* XLV (2): 195–215.

Murove, Felix Munyaradzi. 2005. The theory of self-interest in modern economic discourse: A critical study in light of African humanism and process philosophical anthropology. Ph. D. Dissertation, University of South Africa.

Mwakikagile, Godfrey. 2006. *Tanzania under Mwalimu Nyerere: Reflections on an African statesman.* Dar es Salaam: New African Press.

Mwakikagile, Godfrey. 2009. *Nyerere and Africa: End of an era.* Pretoria: New Africa Press.

Nagel, Thomas. 2003. Rawls and liberalism. In *The Cambridge companion to Rawls,* ed. Samuel Feeman, 1–61. New York: Cambridge University Press.

Ndaba, W. J. 1994. *Ubuntu in comparison to Western philosophies.* Pretoria: Ubuntu School of Philosophy.

Nkrumah, Kwame. 1965. *Consciencism.* New York: Monthly Review Press.

Noddings, Nel. 1986. *Caring: A feminine approach to ethics and moral education.* Berkeley: University of California Press.

Noddings, Nel. 2002. *Starting at home: Caring and social policy.* Berkeley: University of California Press.

Nozick, Robert. 1974. *Anarchy, state and Utopia.* New York: Basic Books.

Ntibagirirwa, S. 1999. A retrieval of Aristotelian virtue ethics in African social and political humanism: A communitarian perspective. M.A. Dissertation, University of Natal.

Nussbaum, Martha C. 2001. *The fragility of goodness: Luck and ethics in Greek tragedy and philosophy.* New York: Cambridge University Press.

Nussbaum, Barbara. 2003. Ubuntu: Reflections of a South African on our common humanity. *Reflections: The Society for Organizational Learning and the Massachusetts Institute of Technology* 4:21–26.

Nyerere, Julius K. 1968. *Ujamaa: Essays on socialism.* Dar es Salaam: Oxford University Press.

Nyerere, Julius K. 1973. *Freedom and development: A selection from writings and speeches 1968– 1973.* Dar es Salaam: Oxford University Press.

O'Keeffe, Janet. 1994. The right to health care and health care reform. In *Health care reform: A human rights approach,* 35–64. Washington, D.C.: Georgetown University Press.

Onah, Godfrey. The meaning of peace in African traditional religion and culture. http://www.afrikaworld.net/afrel/goddionah.htm. Accessed 4 Sept 2012.

Osuji, Peter. 2013. The contribution of African traditional medicine for a model of relational autonomy in informed consent, Abstract. http://digital.library.duq.edu/cdm-etd/item_viewer.php?CISOROOT=/etd&CISOPTR=162271&CISOBOX=1&REC=2.

Pal, S. K. 2002, March10. Complementary and alternative medicine: An overview. *Current Science* 82 (5): 518–524.

Paris, Peter J. 1995. *The spirituality of African peoples: The search for a common moral discourse.* Minneapolis: Fortress Press.

Parratt, John, ed. 1987. *A reader in African christian theology.* London: Latimer Trend and Company Ltd.

Parrinder, G. 1967. *African mythology.* London: Paul Hamlyn.

Petrini, Carlo, and Sabina Gainotti. 2008, August. A personalist approach to public-health ethics. *Bulletin of the World Health Organization* 86 (8). http://www.who.int/bulletin/volumes/86/8/08-051193.pdf. Accessed 5 July 2012.

Petrini, Carlo. 2010, January 18. Theoretical models and operational frameworks in public. *International Journal of Environmental Research and Public Health Ethics* 7:189–202. www.mdpi.com/journal/ijerph. Accessed 8 Aug 2012.

Physicians' Working Group for Single-Payer National health Insurance. 2003. Proposal of the physicians' working group for single-payer national health insurance. 798. http://www.pnhp.org/PDF_files/Physicians%20ProposalJAMA.pdf. Accessed 17 June 2012.

Pontifical Council for Justice and Peace. 2004, October 25. Compendium of the social doctrine of the church. Vatican web site. http://www.vatican.va/roman_curia/pontifical_councils/justpeace/documents/rc_pc_justpeace_doc_20060526_compendio-dott-soc_en.html. Accessed 15 April 2012.

Pope Benedict XVI. 2009, July 7. Caritas in veritate. Encyclical letter on charity in truth. Vatican web site. http://www.vatican.va/holy_father/benedict_xvi/encyclicals/documents/hf_ben-xvi_enc_20090629_caritas-in-veritate_en.html. Accessed 8 July 2012.

Pope John Paul II. 1979, March 4. Redemptor hominis. Encyclical letter on the redeemer of man. Vatican web site. http://www.vatican.va/holy_father/john_paul_ii/encyclicals/documents/hf_jp-ii_enc_04031979_redemptor-hominis_en.html. Accessed 7 Aug 2012.

Pope John Paul II. 1987, December 30. Sollicitudo rei Socialis. Encyclical letter for the twentieth anniversary of Populorum Progressio. Vatican web site. http://www.vatican.va/holy_father/john_paul_ii/encyclicals/documents/hf_jp-ii_enc_30121987_sollicitudo-rei-socialis_en.html. Accessed 7 Aug 2012.

Pope John Paul II. 1991, May 1. Centesimus annus. Encyclical letter on the hundredth anniversary of rerum novarum. Vatican web site. http://www.vatican.va/holy_father/john_paul_ii/encyclicals/documents/hf_jp-ii_enc_01051991_centesimus-annus_en.html. Accessed 7 Aug 2012.

Pope John Paul II. 1993, August 6. Veritatis Splendor. Encyclical letter on the splendor of truth. Vatican web site. http://www.vatican.va/holy_father/john_paul_ii/encyclicals/documents/hf_jp-ii_enc_06081993_veritatis-splendor_en.html. Accessed 8 Aug 2012.

Pope John XXIII. 1961, May 15. *Mater et Magistra. Encyclical letter on Christianity and social progress.* Trans: William Gibbons and Committee of Catholic Scholars. New York: Paulist Press.

Pope John XXIII. 1963, April 11. Pacem in Terris. Encyclical letter on establishing universal peace in truth, justice, charity, and liberty. Vatican web site. http://www.vatican.va/holy_father/john_xxiii/encyclicals/documents/hf_j-xxiii_enc_11041963_pacem_en.html. Accessed 7 July 2012.

Pope Paul VI. 1967, March 26. Populorum progressio. Encyclical letter on the progress of peoples. Vatican web site. http://www.vatican.va/holy_father/paul_vi/encyclicals/documents/hf_p-vi_enc_26031967_populorum_en.html. Accessed 11 April 2012.

Pope Paul VI. 1975, December 8. Evangelii Nuntiandi. Encyclical letter on announcing the Gospel. Vatican web site. http://www.vatican.va/holy_father/paul_vi/apost_exhortations/documents/hf_p-vi_exh_19751208_evangelii-nuntiandi_en.html. Accessed 17 July.

Pope Pius XI. 1931, May 15. Quadragesimo Anno. Encyclical letter on reconstruction of the social order. Vatican web site. http://www.vatican.va/holy_father/pius_xi/encyclicals/documents/hf_p-xi_enc_19310515_quadragesimo-anno_en.html. Accessed 11 May 2012.

Prokes, Mary Timothy. 1996. *Toward a theology of the body*. Grand Rapids: Eerdmans.

Ramose, M. 2002. *African philosophy through Ubuntu*. Harare: Mond Books.

Rawlison, Mary C., and Anne Donchin. 2005. The quest for universality: Reflections on the universal draft declaration on bioethics and human rights. *Developing World Bioethics* 5 (3): 1471–8731.

Rawls, John. 1971. *A theory of justice*. Cambridge: Belknap Press.

Revel, Michael. 2009. Respect for cultural diversity and pluralism. In *The UNESCO universal declaration on bioethics and human rights: Background, principles and application,* eds. Henk A. M. J. ten Have and Michele S. Jean, 199–209. Paris: United Nations Educational and Cultural Organization.

Rhodes, Margaret, P. Battin, and Anita Silvers, eds. 2012. *Medicine and social justice: Essays on the distribution of health care*. New York: Oxford University Press.

Richards, Dona. 1980. European mythology: The ideology of 'progress'. In *Contemporary black thought: Alternative analyses in social and behavioral science*, ed. M. Asante and A. Vandi, 59–79. Beverly Hills: Sage.

Rivard, Glenn. 2009. Non-Discrimination and Non-Stigmatization. In *The UNESCO universal declaration on bioethics and human rights: Background, principles and application,* ed. Henk A. M. J. ten Have and Michele S. Jean, 188–198. Paris: United Nations Educational and Cultural Organization.

Robinson, Fiona. 1999. *Globalizing care: Ethics, feminist theory, and international relations*. Boulder: Westview Press.

Ruch, E. A. 1975. Towards a theory of African knowledge. In *Philosophy in the African context*, ed. D. S. Georgiades and I. G. Delvare. Johannesburg: University of Witwatersrand Press.

Ryan, C. C. 1976. Yours, mine, and ours: Property rights and individual liberty. *Ethics* 87:126–141.

Samuel, Julie, Bartha Maria Knoppers, and Denise Avard. 2011. Medical research involving children: A review of international policy statements. In *The Sage handbook of health care ethics*, ed. Ruth Chadwick, Henk ten Have, and Eric M. Meslin, 261–277. Thousand Oaks: Sage.

Scheman, Naomi. 1983. Individualism and the objects of psychology. In *Discovering reality*, 225–244. Boston: D. Reidel Publishing Company.

Schoen, Cathy, et al. 2005, June 14. Insured but not protected: How many adults are underinsured? *Health Affairs 2005 Web Exclusive* 5:289–302.

Schrecker, Ted. 2008, August. Denaturalizing scarcity: A strategy of enquiry for public health ethics. *Bulletin of the World Health Organization* 86 (8): 600–605.

Schuklenk, Udo, and Cheryl Cline. 2013. Global health ethics. *The international encyclopedia of ethics*, ed. Hugh LaFollette. Blackwell Publishing Ltd.

Second Vatican Council. 1965, December 7. Gaudium et Spes. Pastoral constitution on the Church in the modern world. Vatican web site. http://www.vatican.va/archive/hist_councils/ii_vatican_council/documents/vat-ii_cons_19651207_gaudium-et-spes_en.html. Accessed 17 March 2012.

Senghor, Léopold S. 1964. *On African socialism*. London: Pall Mall Press.

Serour, Gama I., and Ahmed R. A. Ragab. 2011. Ethics of genetic counseling. In *The Sage handbook of health care ethics,* ed. Ruth Chadwick, Henk ten Have, and Eric M. Meslin, 147–148. Thousand Oaks: Sage.

Sevenhuijsen, Selma. 1998. *Citizenship and the ethics of care: Feminist considerations on justice, morality and politics.* New York: Routledge.

Shetty, Priya. UNESCO guidance on ethics and human rights slammed. SciDev.Net. http://www.scidev.net/contentnews/eng/unesco-guidance-on-ethics-and-human-rights-slammed.cfm. Accessed 12 Feb 2012.

Showalter, J. Stuart. 1999. *Southwick's law of hospital and health care administration.* 3rd ed. MI: Health Administration Press.

Shutte, Augustine. 1993. *Philosophy for Africa.* Milwaukee: Marquette University Press.

Sindima, H. J. 1995. *Africa's agenda. The legacy of liberalism and colonialism in the crisis of African values.* Westport: Greenwood Press.

Slote, Michael. 2007. *The ethics of care and empathy.* New York: Routledge.

Some, Malidoma Patrice. 1998. *The healing wisdom of Africa: Finding life purpose through nature, ritual, and community.* New York: Penguin Putnam Inc.

Stengel, Richard. 2009. *Mandela's way: Fifteen lessons on life, love, and courage.* New York: Crown.

Sullivan, Dennis, and Larry Tifft. 2008. *Handbook of restorative justice.* New York: Routledge.

Sultz, Harry A., and Kristina M. Young. 1997. *Healthcare USA: Understanding its organization and delivery.* Gaithersburg: Aspen.

Sundman, Per. 1996. Human rights, justification, and Christian ethics. Ph. D. Dissertation, Uppsala University.

Sweet, William, and Joseph Masciulli. 2011. Biotechnologies and human dignity. *Bulletin of Science, Technology and Science* 31 (1). http://bst.sagepub.com/content/31/1/6. Accessed 8 Aug 2012.

Tandon, P. N. 2009. Protection of the environment, the biosphere and biodiversity. In *The UNESCO Universal Declaration on Bioethics and Human Rights: Background, Principles and Applications,* ed. Henk A. M. J. ten Have and Jean S. Michele, 247–254. Paris: United Nations Educational and Cultural Organization.

Tangwa, Godfrey B. 2010. *Elements of African bioethics in a western frame.* Mankon: Langaa research and Publishing Common Initiative Group.

Taylor, Charles. 1989. *Sources of the self: The making of the modern identity.* Cambridge: Havard University Press.

Taylor, Allyn L. 1999. Globalization and biotechnology: UNESCO and an international strategy to advance human rights and public health. *American Journal of Law and Medicine* 25:475–541.

Taylor, Allyn L. 2002. Global governance, international health law and WHO: Looking towards the future. *Bulletin of the World Health Organization* 80 (12): 975–980. http://www.who.int/bulletin/archives/80(12)975.pdf. Accessed 6 June 2012.

Teffo, Joe. 1994. *The concept of Ubuntu as a cohesive moral value.* Pretoria: Ubuntu School of Philosophy.

Teilhard de Chardin, Pierre. 1959. *The phenomenon of man.* London: Collins.

Teilhard de Chardin, Pierre. 1969. *Human energy.* London: William Collins Sons and Co. Ltd.

Tempels, Placide. 1946. *Bantu philosophy.* Trans: C. King. Paris: Presence Africaine.

Tempels, Placide. 1959. *Bantu philosophy.* Paris: Presence Africaine.

ten Have, Henk. 2011. Foundationalism and principles. In *The Sage handbook of health care ethics,* eds. Ruth Chadwick, Henk ten Have, and Eric M. Meslin, 20–30. Thousand Oaks: Sage.

ten Have, Henk. 2012, March. Potter's notion of bioethics. *Kennedy Institute of Ethics Journal* 22 (1): 59–82.

ten Have, Henk. 2013, September. Global bioethics: Transnational experiences and islamic bioethics. *Zygon* 48 (3): 600–617.

ten Have, Henk A. M. J., and Jean S. Michele. ed. 2009. *The UNESCO universal declaration on bioethics and human rights: Background, principles and application,* 33–34. Paris: United Nations Educational and Cultural Organization.

Tronto, Joan. 1993. *Moral boundaries: A political argument for an ethic of care*. New York: Chapman and Hall.

Tutu, Desmond. 1999. *No future without forgiveness*. New York: Doubleday.

United States Catholic Conference. 2009. *Economic justice for all: Pastoral letter on catholic social teaching and the U.S. economy*. Washington, D.C.: United States Catholic Conference, Inc. http://www.usccb.org/upload/economic_justice_for_all.pdf. Accessed 18 June 2012.

Van der Ven, Johannes A. 1998. *Formation of the moral self*. Grand Rapids: Wm. B. Eerdmans.

Van, Der Marwe, and L. Willie. 1996. Philosophy and the multi-cultural context of (post) apartheid South Africa. *Ethical Perspectives* 3 (2): 1–3.

Vatican II Documents. Vatican web site. http://www.vatican.va/archive/hist_councils/ii_vatican_council/. Accessed 14 May 2012.

Verhoef, Heidi, and Michel Claudine. 1997. Studying morality within the African context. *Journal of Moral Education* 26:389–407.

Villa-Vicencio, Charles, and John De Gruchy. 1994. *Doing ethics in context: South African perspectives*. Maryknoll: Orbis Books.

Villa-Vicencio. 1992. *Charles. A theology of reconstruction: Nation-building and human rights*. New York: Cambridge University Press.

Walters, L. 1986. The ethics of human gene therapy. *Nature* 320:225–227.

Walzer, Michael. 1983. *Spheres of justice: A defense of pluralism and equality*. New York: Basic Books.

Ward, Janie Victoria, McLean Taylor Jill, and Betty Bardige. 1988. *Mapping the moral domain: A contribution of women's thinking to psychology and education*. Cambridge: Harvard University Graduate School of Education.

Washington, Sylvia Ba. 1973. *The concept of negritude in the poetry of Leopold Sedar Senghor*. New Jersey: Princeton University Press.

Webster, D. D. Liberation theology: General information. http://www.globalchristians.org/politics/2/Liberation%20Theology.pdf. Accessed 12 May 2012.

Weil, Simone. 1973. *Waiting for god*. New York: Harper and Row.

Werpahowski, William. 1984. Political liberalism and Christian ethics. *The Thomist* 48 (1): 81–115.

Whitted, Gary. 1993. Private health insurance and employee benefits. *Introduction to Health Care Services*. 4th ed. 332–337. Dover: Delmar.

Widdows, Heather. 2009, Spring. Border disputes across bodies: Exploitation in trafficking for prostitution and egg sale for stem cell research. *International Journal of Feminist Approaches to Bioethics* 2 (1): 5–24.

World Health Organization. 2003. *Shaping the future*. Geneva: WHO.

World Health Organization. 2008. *International recruitment of health personnel: Draft code of practice*. Geneva: WHO.

World Health Organization. 2010a. *Migration of health workers*. Geneva: WHO.

World Health Organization. 2010b. *Guidance on ethics of tuberculosis prevention, care and control*. Geneva: WHO.

World Health Organization. World Health Statistics. 2009. Geneva: WHO.

Wright, R. 1984. *African philosophy: An introduction*. Lanham: University Press of America.

Wynberg, Rachel, Doris Schroeder, and Roger Chennells. 2009. *Indigenous peoples, consent and benefit sharing: Lessons from the San-Hoodia case*. Berlin: Springer.

Zwart, Hub. 2007. Editorial: Statements, declarations, and the problems of ethical expertise. *Genomics, Society and Policy* 3 (1): i–iii, 1–63. http://www.hss.ed.ac.uk/genomics/previousissues/V3N1.pdf. Accessed 17 July 2012.